10

Always On

· · ·

Always On

● ● ●

Language in an
Online and
Mobile World

NAOMI S. BARON

OXFORD
UNIVERSITY PRESS

OXFORD
UNIVERSITY PRESS

Oxford University Press, Inc., publishes works that further
Oxford University's objective of excellence
in research, scholarship, and education.

Oxford   New York
Auckland   Cape Town   Dar es Salaam   Hong Kong   Karachi
Kuala Lumpur   Madrid   Melbourne   Mexico City   Nairobi
New Delhi   Shanghai   Taipei   Toronto

With offices in
Argentina   Austria   Brazil   Chile   Czech Republic   France   Greece
Guatemala   Hungary   Italy   Japan   Poland   Portugal   Singapore
South Korea   Switzerland   Thailand   Turkey   Ukraine   Vietnam

Copyright © 2008 by Naomi S. Baron

Published by Oxford University Press, Inc.
198 Madison Avenue, New York, NY 10016

www.oup.com

First issued as an Oxford University Press paperback, 2010

Library of Congress Cataloging-in-Publication Data
Baron, Naomi S.
Always on : language in an online and mobile world / Naomi S. Baron.
p. cm.
Includes bibliographical references and index.
ISBN 978-0-19-973544-0 (pbk.)
1. Language and languages.   2. Instant messaging.
3. Internet.   I. Baron, Naomi S.   II. Title.
P107.B37 2008
401.4—dc22
2007038058

1 3 5 7 9 8 6 4 2

Printed in the United States of America
on acid-free paper

To the American University Library
(the heart of the university)

Contents

• • • Preface

When Samuel Johnson first set about writing his landmark *Dictionary of the English Language*, he somewhat naïvely believed his task to be setting down, for generations to come, the composition of the English lexicon. More than a decade later, when the long-awaited volumes appeared, Johnson acknowledged his initial folly. In the famous Preface of 1755, he explained that word meanings evolve over time and that the pronunciations of these words do as well. A lexicographer, he wrote, should

> be derided, who being able to produce no example of a nation that has preserved their words and phrases from mutability, shall imagine that his dictionary can embalm language.

When it came to recording pronunciation, once and for all, Johnson was equally adamant about the futility of the task:

> sounds are too volatile and subtile [*sic*] for legal restraints; to enchain syllables, and to lash the wind, are equally the undertakings of pride.

In writing this book, I have felt comparable frustration in attempting to characterize a phenomenon in flux. This time the challenge is not words but technologies and the systems we build upon them for communicating with one another. Those technologies include personal computers and mobile phones, and the systems have such names as email, instant messaging (IM), Facebook, and blogs. An article on IM published in 1998 now reads like quaint history. Statistics collected six months ago are likely out of date.

That said, like Johnson, I am interested in language over the long haul. Terminology (like "social networking sites") may evolve, but the character of the language (or language platforms) has greater shelf life. New forms arise, but more often than not the functions they serve remain surprisingly stable. My ultimate interest in this book is to understand the synergy between technology and language, not to produce a timely data-reference guide. I, like Johnson, attempt to explain, not embalm.

The genesis of this book lies in a pair of questions: What are we, as speakers and writers, doing to our language by virtue of new communication technologies, and how, in turn, do our linguistic practices impact the way we think and the way we relate to other people? I pose these questions by focusing on contemporary language technologies such as IM and mobile phones, though my interest in the interplay between speech, writing, and technology goes back more than twenty-five years. In 2000 I published *Alphabet to Email: How Written English Evolved and Where It's Heading*, which explored the symbiotic relationship between speech and writing in the history of English, with a special emphasis on the role technologies such as the printing press, the telegraph, the telephone, the typewriter, and (jumping nearly a century) email played in that evolution.

With the explosion of online and mobile language in the twenty-first century, the number of people involved in the communication revolution has skyrocketed. In the year 2000, it would have been hard to imagine today's blogosphere or libraries hastening to move their collections online, making them available anywhere, anytime. Although this book mentions email at various points, the work is really about the technologies that followed. In fact, I initially toyed with calling the book *Beyond Email*, a deferential nod to Theodor Geisel's (aka Dr. Seuss's) children's classic *On Beyond Zebra!*, in which Seuss concocts wondrous letters beyond Z. Just so, the web and mobile phone afford us incredible playgrounds for inventing new opportunities for social communication that extend beyond electronic mail.

By training and profession, I am a linguist. In my case, that means that I study language structure and function, how languages are learned and used, how and why they change, and how technology affects all of the above. Since IM, text messaging, and the like are language undergirded by technology, the fit seems logical enough. But there's more to the story of what it means to be a linguist and more to my research agenda.

In the early twentieth century, a tradition arose in America whereby language was to be studied but not judged. The anthropologist Franz Boas inspired this approach, working to combat late nineteenth-century beliefs about primitive peoples who purportedly spoke primitive languages. This was the age of empire (along with America's westward expansion), and what better justification could there be for grabbing land and natural resources from "primitives" than that the European was "civilizing" the barely human natives. Boaz devoted his life to demonstrating how sophisticated the languages and cultures of Native Americans were. His insistence that all languages are equally powerful remains a tenet of the linguistics profession to this day.

What is the problem? Many in the profession find it unseemly to say anything judgmental about language—despite the fact that members of language communities evaluate language all the time. As parents, we convey linguistic judgments to our children when we expand upon their fore-shortened and grammatically skewed "Kitty felled" with the adult version "The kitty fell down." A major point of formal schooling is to establish spoken and written norms by critiquing what students say and write. In the sixteenth century, poets such as John Skelton (in *The Boke of Phyllyp Spar-owe*) bemoaned the "rustye" and "cankered" character of English in comparison to French or Latin. In the ensuing decades, Shakespeare added hundreds of coinages to the English lexicon—not just because he was creative but because the language needed new words to talk about the contemporary world in which Englishmen found themselves. In short, when language is found wanting, speakers and writers oftentimes do something about the problem.

This is not a book about raising children or teaching college writing, and to set the record straight from the outset, I am neither a fusty grammatical prescriptivist nor, as a radio host once introduced me, a curmudgeon. However, my concerns with online and mobile language go beyond the descriptive. I want to understand today's language usage in light of the larger cultural context of literacy and, more specifically, the print culture that emerged in the English-speaking world by the eighteenth century. If it turns out that electronically-based language is altering linguistic norms and expectations, then it seems self-evidently important to understand the nature of those changes and their potential impact on our linguistic and social lives. We may then choose to judge—and even act. But judgment and action (which are not the agenda of this book) are haphazard at best without an understanding of what that language looks like and an assessment of what its impact might be.

Always On represents my take on these issues. I build my case through argument, by analyzing other people's studies, and by presenting data I have gathered in collaboration with colleagues and students. A few words are in order on the original data.

All of the new studies I report on here—on multitasking, on IM conversations, on away messages in IM, on Facebook, and on mobile phone use—were conducted with college students in the United States. Since I make my professional home within the American university system, this population was of natural interest to me—and readily accessible. But this age cohort is also important demographically. The press has focused much of its attention on teenage use of new language technologies. College, however, is a time of transition, when young people put aside some of their adolescent ways and begin defining themselves as adults. To understand what the next group of thirty- and forty-something users of language technologies might look like, it is vital to understand the emerging practices of this transitional group.

Why only American studies? Because these are the students to whom I had access. I am abundantly aware that young adults in the Philippines or Japan are vastly more experienced with mobile phones than their American counterparts, and that we must not assume blogs serve the same function in Iraq as they do in the United States. Age, gender, educational level, and, most important, culture are all potentially critical variables in mapping online and mobile language behavior. In the discussion of mobile phones in chapter 7, I make a few comparative remarks, anticipating the cross-cultural study of mobile phone use in Sweden, the United States, Italy, and Japan in which I am presently engaged.

The last caveat concerns research methodology. Most of the studies involved a combination of quantitative and qualitative data-gathering. Because the number of subjects involved in each project was small, and always constituted a convenience sample rather than a randomly selected group of college students, the studies are best viewed as pilot tests, not statistically elegant research. But the methodology I followed afforded me what I was looking for: a first glimpse at a wide range of language technologies in action. My hope is that more extensive analyses (with far more subjects and more rigorous sampling procedures) will benefit from the new ground this first set of investigations has broken.

• • •

Were I only permitted to acknowledge one person or group that made this book possible, thanks would go to the students I taught at American University between 2001 and 2005 in my University Honors Colloquium, "Language in the New Millennium." These students helped me in collecting (and, in some cases, analyzing) data, but equally importantly, they gave me a forum through which to develop many of the conceptual frameworks presented here. I am grateful to Michael Mass, director of the University Honors Program, for inviting me to teach this course in its multiple iterations.

Among my students, special mention goes to Lauren Squires, Sara Tench, and Marshall Thompson for their assistance with the study of IM away messages (see chapter 5) and of IM conversations (see chapter 4). Tim Clem and Brian Rabinovitz played key roles in constructing and administering the study of multitasking while doing IM (see chapter 3). Over the years, Lauren and Tim, along with Erin Watkins, have been invaluable sources of insight regarding the world of electronically-mediated communication. Erin is the one who first introduced me to Facebook—a debt I attempt to repay through my analysis in chapter 5. I am also grateful to Clare Park for assistance in collecting Facebook survey data.

Professionally, Rich Ling has been my invaluable "go-to" colleague for mobile communication. Since 2001, he has taught me a good deal about mobile phone issues, and I have been fortunate to collaborate with him on a survey of American college-student mobile phone practices and a comparison of American text messaging versus American IM (see chapter 7). For the survey, I am also indebted to Katie Young, Laura Deal, and Gia DiMarco, who did a preliminary analysis of some of the data.

In 2001 Americans were already avid users of email (and to a lesser extent IM), but they knew precious little about mobile phones. Yes, a growing number of Americans had begun buying phones, but they were not yet widespread, and text messaging was largely unheard of. Serendipitously, I attended a conference in spring 2001, run by James Katz at Rutgers, called "Machines that Become Us." I came away feeling like Alice after stepping through the rabbit hole, having discovered a world of communication possibilities, along with a host of new colleagues.

My gratitude also to the Association of Internet Researchers (AOIR). Though itself a relative newbie (its first conference was in fall 2000), AOIR has become an invaluable forum for exchanging ideas about the Internet and all that it entails. AOIR's conferences are true oases for scholars looking for serious discussion of topics too new for many university catalogues. Equally vital is the listserv run by AOIR, which creates a year-round network of scholarly exchange, without which this book would have been the poorer.

My thinking has been clarified though the collegial give-and-take that occurred by presenting my work at conferences and in lectures. Special thanks to Rich Ling (2003 in Grimstad, Norway), Santiago Postaguillo (2003 in Castellón de la Plana, Spain), Joe Walther (2004 in New Orleans, Louisiana), Mark Aronoff (2005 in Washington, DC), Jim Katz (2006 in New Brunswick, New Jersey), Leopoldina Fortunati and Maria Bortoluzzi (2006 in Pordenone and Udine, Italy), Andreas Jucker (2007 in Zurich, Switzerland), and Gerard Goggin and Larissa Hjorth (2007 in Sydney, Australia) for invitations to work with splendid colleagues and students.

In the process of writing books, authors accrue debts to colleagues whose contributions are sometimes less concrete but equally invaluable. Through their critiques of my work, sharing of their own research, and willingness to educate me, such people have made this book stronger. My list (in alphabetical order) includes Jack Child, David Crystal, Brenda Danet, Leopoldina Fortunati, Ylva Hård af Segerstad, Susan Herring, Mizuko Ito, Amanda Lenhart, Rich Ling, Misa Matsuda, Anabel Quan-Haase, Sali Tagliamonte, Crispin Thurlow, Marta Torres, Barry Wellman, and Simeon Yates.

Always On also has benefited from feedback given by named (or anonymous) readers of earlier papers and book chapters I have written dealing with online and mobile communication. Rich Ling, Beth Scudder, and Karen Taylor have kindly read the manuscript, offering insightful advice (and at a number of points saving me from my own folly). Any remaining errors of fact or judgment are, of course, my own fault.

Paul Budde Communication Pty Ltd provided helpful statistics on global Internet usage. For permission to reproduce the cartoons appearing in the text, I am grateful to the Cartoon Bank, United Media, and Creators Syndicate. Special thanks to William Hamilton for redrawing his cartoon when the original was not findable, and to Don Wright for permission to reproduce his cartoon. Diane Rehm graciously gave me a copy of the commencement address she delivered at American University, from which I quote in chapter 10. The Faculty Corner at American University's Center for Teaching Excellence provided timely support with technology at many points along the way. Kevin Grasty, director of University Publications at AU, offered much-appreciated graphic assistance.

Working with Oxford University Press has been a pleasure. My editor, Peter Ohlin, has been a source of wise counsel throughout. Molly Wagener and Joellyn Ausanka have rendered the production process both efficient and enjoyable.

As always, my family has been my mainstay in the authorial journey. Nikhil and Leslie stoically endured another of my book ventures. My son Aneil was equally supportive at every turn, including his early willingness to Friend me on Facebook. Special gratitude also goes to Anne Wutchiett, my yoga and pranayama teacher at Unity Woods, who has helped keep me centered, healthy, and breathing.

For two decades, American University has been my intellectual home, affording me opportunities to try out ideas in a collegial environment. Through sabbatical support in 2000–2001, I was able to begin formulating some of the arguments that would make their way into the present volume. Travel allocations through the College of Arts and Sciences, including the CAS Mellon Fund, made it possible for me to present aspects of my work at national and international conferences. Reduced teaching loads have afforded me the luxury of devoting precious time to research and writing during the academic year.

My special gratitude goes to the American University Library, which is the unsung campus hero. While the physical collection is not vast, its librarians have unstintingly provided the best services to faculty and students I have encountered anywhere in my university career. It is therefore to the AU Library that this book is dedicated.

Always On

• • •

1

• • • Email to Your Brain

Language in an Online and Mobile World

On a warm Texas afternoon, a young boy walked lazily down the middle of a dirt road. On either side, barbed wire fenced off fields for crops or cattle. Suddenly, a roar came out of nowhere, dust-clouds swirled, and a black chimera lurched straight toward him. Terrified, the boy dived for the side of the road, cutting himself badly on the barbed wire.

The year was about 1905. The youth had just seen his first automobile. This story was recounted to me by Billy Fullingham, a colleague at Southwestern University in Georgetown, Texas, where I was a visiting professor in the mid-1980s. The young man was her father.

Fast forward to the early twenty-first century and another Billy—this time, the Reverend Billy Graham. The occasion was one of the last of Graham's legendary revivals. Always a man to reach out to his listeners, to speak in words that would resonate, Graham proclaimed, "Conscience is the email God sends to your brain."

• • •

These two incidents, though separated in time, illustrate our very human reactions to technology. When technology is new (as in the case of the automobile), bewilderment—even fear—is a natural response. When a technology has become embedded in our everyday practices (as with email), a metaphor such as "email to your brain" is understood effortlessly. Roger Silverstone and Leslie Haddon have used the term "domestication" to describe the process whereby a new device (such as a car, a vacuum cleaner, or a mobile phone) becomes a normal component of daily living.[1]

Take another example of domestication of a technology: flying on an airplane. In the early decades of commercial flights, traveling by plane was a formal event. People dressed for the occasion: gloves and hats for ladies, suits for gentlemen, shined shoes for all. Today, however, flying is thoroughly domesticated. Unless they are heading directly to a business meeting, passengers

3

may well wear jeans or even sweatpants, making themselves comfortable for the long haul ahead.

Since this book is about communication technologies, our question regarding domestication is this: How does our language evolve, along with changes in the way we interact with other people, as communication technologies become increasingly domesticated? Of course, domestication is not an all-or-none proposition. Ten years ago, many Americans were daily users of instant messaging, but much of the rest of the country had not heard of IM. Those innocent of one language tool might have domesticated a different medium. You are an IM devotee, while I relentlessly check email. You live by your BlackBerry, while I am inseparable from my mobile phone.

Our cumulative experience with communication technologies has gradually altered behavioral and social norms. It is easier to find the weather forecast online than to tune in to the Weather Channel on television and await the local report. Email or a phone call (sometimes resulting in a voicemail) replaces walking from our office to the one next door to ask a question or deliver a message. As domestication of communication technologies spreads through the populace, people are increasingly "on" networks that extend beyond the landline phone.

We commonly speak of universal access to landline telephones in developed parts of the world. Literally, of course, not everyone who might want a telephone has one. But at least in the United States, the proportions are large enough that for the sake of discussion we can speak of everyone having landline access—if not to his or her own phone, then to someone else's. Universal access to newer communication technologies using computers or mobile phones has not yet arrived, but is showing impressive growth in most parts of the globe.

Among those people who are "on" modern communication technologies, an increasing number are "always on." Again, "always," like "universal," is a generalization, but one that makes sense here. Not everyone drives a car or flies on airplanes, but essentially we're all familiar with the technologies. Just so, Billy Graham knew that even those among his audience who themselves did not use email were aware of the medium, rendering his metaphor comprehensible.

How has the growing domestication of email, IM, text messaging on mobile phones, blogging, Facebook—and the rash of other forms of online and mobile communication platforms—altered our communication landscape? Some of the effects are obvious. Once you have the requisite equipment (a computer, a mobile phone) and have managed the access fees, it's far simpler and less expensive to communicate with people not physically present than at any time in human history. A second palpable change is the ease with which each of us can become an author or publisher. Whether with blogs, web pages, or emailing files to distribution lists, we can bypass the traditional textual gatekeepers: editors of all ilk (newspaper, periodical, book) and, sometimes, legal authorities. What is more, these new technologies potentially compromise our privacy and even our individual safety. On the annoying-but-comparatively-innocuous end of the spectrum, we sometimes hit "send" on an email before we have finished (or edited) what we meant to say; we accidentally forward messages to the wrong people or intentionally forward missives we received as private communiqués. Much more dangerously, online financial predators ask for social security numbers or banking information, and lonely teenagers are victimized by online "friends."

Other consequences of these electronic language media are less certain. For years, the popular press has been asking whether email, IM, or texting on mobile phones is degrading the way we write. Conversely, some linguists and composition teachers argue that all this writing is bringing about an epistolary renaissance that is strengthening our language abilities—and the language itself. Similarly, there is considerable disagreement over the social consequences of doing so much communicating-at-a-distance rather than face-to-face. One group of researchers suggests an inverse relationship between time spent online and social well-being. Their opponents adduce evidence that heavy online communicators generally have ample social meetings "in real life."

Beyond the obvious—and the contended—effects of new language technologies, there are, I suggest, two fundamental changes that email and its descendants are having upon our language and the way we use it. These two transformations are at once subtle and potentially invidious, challenging our assumptions about interpersonal communication and calling for us to rethink conventional notions about spoken and written language.

The first of these changes involves our growing ability, using communication technology, to assert control over when we interact with whom. Consider the case of a telephone ringing. In the early days, if the phone rang you answered it—regardless of what you were doing or with whom you might be speaking. Much as drivers of automobiles yield to pedestrians, personal or social activity yielded to the telephone. These days, the unwritten rules are different. Caller ID (on both landlines and mobile phones) enables us to screen calls before choosing whether to answer the summons or let the caller go to voicemail. The growing use of distinctive ring tones (keyed to different people in your mobile phone address book) even obviates the need to haul the phone out of your purse or backpack to decide.

Historically, we have always had some options for what I call controlling the "volume" on our social interactions: crossing the street to avoid

"Spare a little eye contact?"

an unwelcome conversation; ignoring a letter that requests a response. As technology multiplies these possibilities, and as social practices begin shifting to exploit these new tools for interpersonal "volume" control, are the social relationships themselves affected?

The second transformation concerns the amount of writing we are now doing and what effect quantity may be having upon quality. A few years back, I did a National Public Radio interview on the possible effects of instant messaging on the speech and writing of teenagers. The other guest, himself a polished writer, extolled the benefits that email and IM seemed destined to have upon the next generation's writing abilities. I countered with the proverbial case of monkeys at typewriters: However long they pound away, they are unlikely to produce Shakespeare.

More recently, I have begun to suspect that the situation is even more troubling. Could it actually be that the more we write online, the *worse* writers we become? I'm not talking about whether the usual litany of abbreviations and acronyms (such as 2 for 'to' or 'two,' or *btw* for 'by the way') is seeping into everyday writing, or whether our emails are laced with misspelled words or minimalist punctuation. These are issues we will address in due course. Rather, my concern is more profound: Is the sheer fact that we are replacing so much of our spoken interaction with written exchanges gradually eroding a public sense that the quality of our writing matters?

I vividly recall a piece from the *Times Literary Supplement* in early 2000, in which the reviewer despaired over the profusion of spelling and punctuation mistakes he had found in the text at hand (which, incidentally, had been published by a highly respected press). Worse still, he noted, this book was not unique. Sardonically, he mused that about ten years earlier, all competent proofreaders must have been taken out and shot. I was reminded of this comment in late 2006, when I came upon a slick full-page ad, placed in a university alumni magazine, for upscale condominiums in California. The ad pictured a successful-looking young woman (who happened to be Caucasian) stepping out of the back seat of a car (presumably a limousine). The disclaimer at the bottom of the ad read, "Model depicted do not reflect racial preference." A politically correct sentiment no doubt, but what about the grammar?

Is the problem actually the proofreaders? Or might it be that we the readers (who ourselves are often writers) are less fussy than we used to be? Is it that we *could* proofread—we know the rules—but no longer care to do so? More radical still is the issue that even good writers are themselves becoming less certain about rules for word construction and sentence mechanics. Is it "iced tea" or "ice tea"? "Ring tone" or "ringtone"? And so what?

In 2003, John McWhorter wrote *Doing Our Own Thing: The Degradation of Language and Music and Why We Should, Like, Care*, in which he argued that contemporary Americans, unlike earlier generations and unlike many other cultures, do not particularly care about their language. In McWhorter's words, "Americans after the 1960s have lived in a country with less pride in its language than any society in recorded history."[2] While I believe McWhorter is substantially correct, his point is not the one I am making here. My own argument is that the sheer amount of text that literate Americans produce is diminishing our sense of written craftsmanship. To rephrase Thomas Gresham, bad writing is driving good writing out of circulation.

Beyond these two transformations in our use of speech and writing, I have begun to sense a third effect of language technologies that is much harder to articulate but which stands to reshape how we interact socially with one another. To the extent language technologies make it possible to always be in contact, we end up sharing a great deal of information and experiences, which in earlier times we might have saved up for face-to-face meetings. Children at summer camp IM their friends back home, and camp administrators post photos of the day's adventures on web sites for parents to access. College students use mobile phones to call home, sometimes daily, offering play-by-play accounts of their activities and angst.

I have taken to calling this phenomenon "the end of anticipation," because we no longer await the return of family and friends to share in their stories. For as long as humanity can remember, anticipation of reunion has been part of our social definition. That is, a relationship is a composite of joint experiences plus recounting of events taking place while we are apart. Letter-writing of old afforded a selective (and often reflective) window on our activities, but the contents of these letters were then often embellished during later physical encounters. If we are always together virtually, we may need to redefine the substance of meeting again face-to-face. I wouldn't for a moment want to forego viewing the photo albums my son posted on Facebook when he was studying in Paris. But that near-real-time window on his world restructures our subsequent time together in new ways that we need to understand.

• • • EXPLORING LANGUAGE TECHNOLOGIES: WHAT'S IN *ALWAYS ON?*

The book begins with an overview of what we mean by language in an online and mobile world (chapter 2). The chapter offers a chronological précis of the types of language media to emerge over the past thirty-five years, with particular emphasis on some of the newer communication platforms. A bit of terminology is introduced to give us a common vocabulary.

Chapter 3 develops the first major theme of the book: using language technology to control the volume of interpersonal communication. The analysis examines a range of language technologies that have increasingly empowered us to call the shots on social interaction. The more conceptual discussion is illustrated with data from a study my students and I did in fall 2004 and spring 2005 of multitasking behavior by college students while using instant messaging. The study reveals how deft the current generation of multitaskers is at meshing online conversations with other activities.

Each of the next four chapters focuses on a particular genre of online or mobile language. Chapter 4 looks at the linguistic guts of IM conversations. My data were collected in spring 2003, though more recent studies of IMing by teenagers and young adults confirm many of the findings. The analysis here is more detailed (and has more terminology) than any other part of the book. However, given the question I'm trying to answer—Is IM just informal speech written down?—I need to draw upon a larger linguistic toolkit.

Chapter 5 moves us from one-to-one communication to the way people present themselves to a select group of friends online. The data here come from two sources: a study of IM away messages that my students collected in fall 2002, coupled with a spring 2006 study of how college students were using and responding to Facebook. Though the platforms differ markedly, student behavior patterns are surprisingly similar.

In Chapter 6 we move the boundary markers for audience even farther out, looking at blogs, YouTube, and Wikipedia. The chapter reviews the emergence of these new social media, but the crux of the discussion focuses on why these technologies have experienced such meteoric growth. At least part of the answer seems to be that new media are essentially substitutes for earlier outlets of public expression, including newspaper letters to the editor and talk radio.

All of the language technologies discussed in chapters 4, 5, and 6 are historically computer based. Chapter 7 turns to mobile telephones. Before proceeding any further, a terminological clarification is in order. In North America, the term of choice is "cell phone," essentially referring to the fact that telephone signals are passed from one zone (or "cell") to another. Practically the entire rest of the world calls those same instruments "mobile phones," highlighting the portability of the devices themselves. In this book, I have chosen to use the term "mobile." It is the mobility of phones (and other portable communication devices) that I want to talk about, not the details of transmitting electrical signals. What's more, settling upon a single term helps avoid schizophrenia when talking about statistics and usage patterns around the world.

While mobile phones have been well entrenched in Europe and Asia for at least a decade, America's love affair with mobiles is more recent, especially when it comes to pecking out text messages. After setting some historical context, the chapter focuses on two studies I conducted in fall 2005 with my colleague Rich Ling (of Telenor in Norway), along with the invaluable assistance of my students. The first project assessed how American college students use—and feel about the use of—mobile phones, while the second investigation compared the linguistic structure of American college-student text messages with the IM conversations I gathered in spring 2003.

Moving from particular types of online and mobile language to more general linguistic issues, chapter 8 takes on the question I have repeatedly been asked by the news media over the past decade: Is the Internet destroying language? The chapter attempts to answer this question by drawing upon the empirical data presented in the book thus far, as well as by working through some of the conceptual and social assumptions underlying linguistic practices today.

The final two chapters lead us from microcosm to macrocosm, putting what we have learned about online and mobile language into broader linguistic and social perspective. Chapter 9 asks, Whither written culture? What assumptions regarding reading and writing, in effect for nearly three hundred years, are being challenged by computers and mobile phones? Given all the writing we are now doing in lieu of face-to-face or telephone conversations, are we flooding the scriptorium, causing us to devalue the writing we are producing and reading?

Beyond the effects that contemporary media may be having upon our language, we need to think about whether computers and mobile phones are impacting the social fabric as well (chapter 10). Since the early days of mainframes, many people have feared that computers are undermining our sense of community. These concerns proliferated with the explosive growth of computer-based communication such as email. The good news is that most contemporary studies examining the social effects of Internet use indicate we have more cause for relief than concern.

Even if avid email users are not doomed to be social recluses, there are subtle—and perhaps more troubling—ways in which communicating online (or by mobile phone) is reshaping us, less by virtue of the mechanisms themselves than by the way we use them. Increasingly, more and more people are "always on" one technology or another, whether for communicating, doing work, or relaxing by surfing the web or playing games. Regardless of the purpose, the fact that we are always on means that we need either to drop some other activity or multitask. And so our final question in the book is this: What kind of people do we become—as individuals and as family members or friends—if our thoughts and our social relationships must increasingly compete for our attention with digital media? These are not simply academic questions for scholars to debate at conferences. Rather, the answers directly affect each one of us.

• • • HOW TO READ THIS BOOK

Always On is written for a variety of readers: people curious about the Internet and mobile phones, teachers and parents trying to get a fix on the likes of IM and blogging, students of new media, linguists seeking a scholarly analysis of online language. Writing for all of these audiences at once can be a challenge. Some of the topics we deal with lend themselves to close reading while the tone of others is more conversational.

The book is designed to be read from start to finish, but depending upon your interests you may gravitate to some chapters more than others. Results from my empirical studies appear in chapters 3, 4, 5, and 7. People wanting just "the big picture" can figure out what to skim in the relevant sections.

If you are one of those people who jump to the end of mystery stories to discover "who done it," then plunge forward to chapters 8, 9, and 10. My hope is that by the time you have read the conclusions, you'll want to go back to the earlier chapters to discover the rationale behind them.

2

• • • Language Online

The Basics

It was early November 1493 when Christopher Columbus and his crew arrived on the island of Guadeloupe in the Caribbean. Among his discoveries was a strange fruit, known locally as *nana*. Writing in his diary, Columbus explained that the fruit "is shaped like a pine cone, but it is twice as large and its flavor excellent. It can be cut with a knife like a turnip, and it seems very healthful."[1]

Columbus brought the fruit back with him to Spain, from whence it made its way to England in the mid 1660s. But what was the new delicacy to be called? The simple solution was to piece together words already in the language: *pine* (because the base resembled a pine cone) and *apple* (at the time, still the generic term for "fruit").

As technology has evolved, new devices have often been named (at least temporarily) by familiar words and concepts. The telephone was originally designed as a "harmonic telegraph." What today we call *movies* were first known as "talking pictures" or "talkies." Other original terms have stuck: An automobile (that is, *auto*, as in "automatic," plus *mobile*) is still *automobile*—unless, of course, it is *car*.

As the functions of computers expanded from doing computations to storing data, creating documents, and enabling people to communicate through networking, again there was a need for new nomenclature. In the early days of cross-machine communication, a number of terms began appearing in the nascent literature to denote language appearing online: "interactive written discourse," "e-mail style," or "electronic language." A few years ago, David Crystal introduced the word "Netspeak," denoting the linguistic features characterizing the range of Internet-based language.[2]

In the 1980s the term "computer-mediated communication," more commonly known as CMC, emerged to encompass a range of platforms used for conversing online, including email, listservs, chat, or instant messaging. With the development of mobile devices such as the BlackBerry and mobile phones, which aren't really computers, the term CMC became something of a stretch. Many researchers began speaking of information communication technologies (ICTs), alluding to the machines themselves (computers, personal digital assistants, mobile phones) rather than to the information they conveyed. What we now needed was an umbrella term for various types of language transmitted via the gamut of ICTs. Several colleagues and I have begun speaking of "electronically-mediated communication" (or EMC) because, like *pineapple*, the phrase does its job.

So much for terminology. What kinds of online and mobile language are we talking about?

• • • IN THE BEGINNING

The origins of new technologies often turn out to be more prosaic than popular imaginings. We understandably assume that Alexander Graham Bell's famous 1876 call to his assistant, "Mr. Watson, come here. I want you!" signaled the scientific triumph of conveying the first voice message across a telephone line. In actuality, Bell summoned Thomas Watson from the next room because he, Bell, clumsy as usual, had spilled sulfuric acid on himself and needed help cleaning up.[3]

The first email message was equally mundane. In 1971 Ray Tomlinson (a computer engineer working at Bolt Beranek and Newman) sent an arbitrary string of letters between two minicomputers that, although networked through a precursor of the Internet, were actually sitting in the same room. This first email was hardly an exercise in interpersonal communication. It did, however, engender a convention that helped define the way all email henceforth would be sent. To clarify the recipient and machine location to which a message was addressed, Tomlinson selected the @ symbol, which separated a user's login name from the name of his or her computer.[4] Today, we are all too familiar with the format of email addresses such as jcaesar@rubicon.mil.

Telephones and email are just two of the technologies for communicating information across distances. Since the human voice can reach only so far, societies have long used smoke signals or drum beats to convey messages to those outside of earshot. Semaphores and the telegraph were more sophisticated technologies for accomplishing the same goal. With the development of computers, written messages could be transmitted only if there were a system for linking machines together. Therefore, our story of electronically-mediated communication begins with a brief look at the networking systems that made EMC possible.

The earliest computer networks were created by the U.S. military for sharing numerical data between research sites. Over time, the same binary coding system developed for sending numbers was used for transmitting language. ARPANET (the U.S. Department of Defense's Advanced Research Projects Agency Network) was built between 1968 and 1969, under a contract with Bolt Beranek and Newman.

Civilians began joining the networking community in the late 1970s and early 1980s. Homegrown bulletin board systems (BBSs), carried over telephone dial-up lines, connected clusters of friends and helped create the earliest online social communities. While the best known of these groups was the WELL (Whole Earth 'Lectronic Link), the number of online communities quickly mushroomed. Computer scientists were not far behind in creating networking systems that were independent of the military-based ARPANET. In 1979/80, USENET (UNIX Users Network) was developed at the University of North Carolina as a kind of "poor man's ARPANET."[5] An important function of USENET was to carry distributed online forums known as newsgroups (a form of CMC we'll return to in a moment).

Enter the Internet in 1983. Over time, through a few twists and turns, the old ARPANET became the Internet, which was a federally funded project linking multiple computer networks through a specific type of communication protocol known as TCP/IP.

The infant Internet was a potentially dynamic tool but not one easily harnessed. In the early 1990s, Tim Berners-Lee designed the World Wide Web, essentially a collection of software tools and protocols that make it relatively easy for computers to communicate across the Internet. A number of earlier functions (such as email) were ported to the web, making the exchange of information incredibly smoother.

The most important step toward user-friendliness was the emergence of tools for searching the web. Having thousands of web pages out in cyberspace was of little tangible value if you didn't know where to find them. The 1990s saw the rapid appearance of a succession of search tools, most of which were free to end-users. Gopher (also the name of the University of Minnesota's mascot) was designed in the early 1990s for locating documents on the Internet. In 1993, Marc Andreessen at the University of Illinois created the web browser Mosaic, the commercial version of which, Netscape, appeared in 1994. Microsoft's Internet Explorer followed in July 1995. In September 1998, Google made its debut. By March 2007, roughly 3.8 billion Google searches were being done in the United States per month.[6]

All these networking (and search) tools provide infrastructure for transmitting written language online. But how are the messages conveyed?

• • • SORTING OUT THE OPTIONS

Electronic communication can be divided up along two dimensions. One is synchronicity: Does communication happen in real time (synchronous), or do senders ship off their messages for recipients to open at their convenience (asynchronous)? The other dimension is audience scope: Is the communication intended for a single person (one-to-one) or for a larger audience (one-to-many)? Here's what the scheme looks like:

	asynchronous	*synchronous*
one-to-one	email, texting on mobile phones	instant messaging
one-to-many	newsgroups, listservs, blogs, MySpace, Facebook, YouTube	computer conferencing, MUDs, MOOs, chat, Second Life

In terms of chronological appearance, here's another view of the specific technologies:

1971	Email
1971	Early Computer Conferencing
1979	MUDs (Multi-User Dungeons/Dimensions)
1980	Newsgroups
1986	Listservs
1980s, early 1990s	Early Instant Messaging (IM) (e.g., UNIX talk, ytalk, ntalk)
1988	IRC (Internet Relay Chat)
1990	MOOs (MUDs, Object Oriented)
1992	Text Messaging on Mobile Phones
1996	ICQ ("I Seek You") (modern IM system)
1997	AIM (America Online Instant Messenger)
1997	Blogs (Web Logs)
2003	Second Life
2003	MySpace
2004	Facebook
2005	YouTube

Given the pace at which online language technologies have evolved, it's easy to lose track of the historical roots of today's latest communication platforms. And in many cases, the identity of innovators has become obscured. Our discussion acknowledges how modern electronically-mediated communication builds upon the hard work of its predecessors.

For clarity, I've organized the overview according to the four-way schema of asynchronous versus synchronous, and one-to-one versus one-to-many. In this chapter, we focus on written communication. Later on, we'll look at audio and video exchanges. Some of the technologies we discuss are obviously yesterday's news. They're included here both for historical completeness and to illustrate that contemporary communication tools are often filling earlier functions: New bottles for old wine.

A word of caution: Although it's common to speak of asynchronous versus synchronous communication as if the two are polar opposites, in actuality they fall along a continuum. In a sense, the only real synchronous communication is that in which one person can interrupt another—the prototypes being telephone conversations or face-to-face speech.[7]

One-to-One: Asynchronous

Email

Without question, email became the killer application for networked computers, once the Internet was in place and the cost of computer hardware and connectivity had begun to drop. The technology is now an indispensable part of modern work and play, love and war.

*"Hi. My name is Barry, and I check my E-mail
two to three hundred times a day."*

© *The New Yorker* Collection 2001 David Sipress from
cartoonbank.com. All Rights Reserved.

In principle, email is a one-to-one asynchronous medium. However, neither of these characteristics is always true. Senders and recipients are free to broadcast messages as they see fit, either publicly or sub rosa. Jack may email Jill but send copies (declared or as blind copies) to Tom, Dick, and Harry. In turn, Jill may take email from Jack and forward it to Jane, along with the new subject line "What an Idiot!" The idea of synchronicity is also up for interpretation. Twenty years ago, it sometimes took hours (even days) for an email to wend its way from me to you. These days, computer servers and signal transport speeds have improved enormously. Lag time may be as short as a second or two, making email essentially synchronous, if you choose to use it that way.

Much has been written about email, but curiously, we have very little tangible data beyond anecdotes.[8] Researchers are often hesitant to ask colleagues—or strangers—for logs of their email correspondence, perhaps for fear people will say no. As a result, the majority of empirical studies of computer-mediated communication have been of one-to-many public forums such as newsgroups and chat, where the researcher can pull quasi-public transcripts off of the Internet.

What we do know definitely about email is that it shows incredible variety in both form and function. In response to parental pressure, a reluctant ten-year-old sends Grandma an email, thanking her for a Christmas present, while a trial lawyer puts opposing counsel on notice, by email, that his client refuses to settle. After dispatching a hastily written email to a friend, saying I'll be late for our luncheon meeting today, I turn around and carefully edit my email requesting a larger budget next year. Trying to characterize email style with a "one size fits all" definition is about as meaningless as describing an "average" American meal: meatloaf or minestrone? potato pancakes or pad thai? cola or cappuccino? By now, email has become sufficiently domesticated, at least in the United States, that its style and content are as diverse as the people using it.

Text Messaging on Mobile Phones

In America, relatively easy access to computers made email, and later instant messaging, convenient ways of sending written communiqués to family, friends, and co-workers. By contrast, in much of the world, especially where computers were less ubiquitous, mobile phones largely assumed these functions.

In the early 1990s, a multinational European effort known as Groupe Spécial Mobile, or GSM, established a uniform mobile telephone system for much of Europe. Over time, and as the GSM network has been adopted by large sections of the globe, GSM has come to mean "Global System for Mobile Telecommunications."

The GSM system was originally designed to convey voice signals from one place to another, much as landline phones do. When the project was essentially complete, a bit of bandwidth was left over. GSM allowed customers to use this space for pecking out simple written messages on the phone keypad. For example, on the "2" key, one short tap would represent the letter *A*; two taps, the letter *B*, and three taps, the letter *C*. Lettering had already appeared on mobile phones, a relic of the days in which area telephone exchanges had names. (When I was a child, my phone number was "GR 4-2525," with the "GR" standing for "Greenbelt," the name of the town with that exchange. Today, the same number would be "474-2525.")

Thus text messaging was born. On the GSM system, texting was known as SMS, standing for Short Message Service. In everyday parlance, most people spoke of SMS as meaning "short text messaging." With time, GSM turned SMS into a highly lucrative business, particularly because the costs per transmission were lower than for voice calls. Teenagers and young adults—whose funds were generally limited—became heavy users of the service, creating an immensely popular mobile language medium in the process.

These days, several alternatives have emerged for simplifying texting input. A number of handset manufacturers offer phones with full tiny keyboards (similar to a BlackBerry). Predictive texting programs (also sometimes known as T9 programs) enable users to type one or two letters of a word, and then a software program offers up the full word, predicting the user's intent.[9] Chapter 7 looks at mobile telephony, both talking and texting, with special focus on how the technology is emerging in the United States.

One-to-One: Synchronous

Instant Messaging

Returning from mobile phones to computers, our next stop on the EMC tour is instant messaging (IM). In principle, the essential difference between email and IM is synchronicity: Email is asynchronous and IM is synchronous. I might send you an email at midnight and not expect a reply until a decent hour the next morning. With IM, I only message you when I know you are online and there is good reason to anticipate a boomerang reply. At least, that is how, in principle, the two systems of one-to-one communication work. Chapters 3 and 4 reveal that the world of IM is actually more nuanced.

The emergence of IM as a communication technology was a two-stage process. Stage 1, which dates to the 1980s and early 1990s, took place on a limited number of American university campuses and research sites, with the development of UNIX applications with names such as "talk," "ytalk," and "ntalk," and the Zephyr IM system created through Project Athena at the Massachusetts Institute of Technology.[10] In the late 1990s, IM become a widespread phenomenon, thanks in large part to Mirabilis Ltd's ICQ ("I Seek You") and to the technology and marketing efforts of America On-line (especially AIM—AOL Instant Messenger). ICQ, which first appeared in 1996, was purchased by AOL in 1998. Other contemporary players in the IM market include Yahoo! Messenger, MSN Messenger, and Google Talk. Increasingly, today's IM systems provide audio and video options as well.

Most IM platforms offer far more than just opportunities for carrying on synchronous conversations. Typical add-ons include profiles, buddy lists, away messages, and the ability to block would-be message-senders.

Profiles are personal information forms, in which users can input contact information (physical addresses, mobile phone numbers, email addresses); date of birth and home town; favorite quotations, books, or bands; and so forth. Often posted in innocence, information on the profile has gotten many teenagers into serious trouble from predators.

Buddy lists are devices for defining your circle of friends. Essentially they are lists of the IM screen names (the IM equivalent of email addresses) of those people with whom you want to share information about your comings and goings. Your buddies know when you are logged on to IM and when you are offline. (Traditionally, you could only send a message when you were visibly online and the person with whom you were IMing was as well, though other options have been introduced more recently.)

Buddies also have access to so-called away messages, which people ostensibly post to announce that although they are still logged on to IM, they won't be checking messages because they have moved away from their computer—to get something to eat, take a nap, go to the bathroom, or attend class. In reality, away messages have become far more creative tools in the hands of teenagers and young adults.

What if you don't want particular people on your buddy list to contact you? The drastic solution is to remove these individuals from your list altogether. A temporary fix is to block a specific person, which can easily be accomplished by changing a single IM setting. When blocked individuals log on to their IM systems, they appear to be offline and therefore not available to be messaged.

One-to-Many: Asynchronous

Newsgroups

The earliest asynchronous form of one-to-many online communication was newsgroups. These public forums (some still exist) originally resided on USENET. Newsgroups entail postings to a common public site, which can be accessed whenever users choose to log on. The network of different newsgroups has historically been vast. Tens of thousands of groups represent seemingly every topic imaginable, from sex to antique cars to medicine. Newsgroups are not restricted in membership. Consequently, the language appearing in posts varies enormously, both in style and propriety.

As the Internet has evolved, the role of newsgroups was largely supplanted by new one-to-many forums, both asynchronous (such as blogs that invite comments) and synchronous (most notably chat rooms). The newsgroups that flourished in the 1980s and early 1990s established important precedents for publicly posted discussions with strangers.

Listservs

If newsgroups fostered conversation with outsiders, listservs were developed to communicate within social groups whose members knew each other or at least worked in the same organization. Listservs descend from mailing list programs created for sharing information across the ARPANET. As the popularity of mailing lists spread in the 1980s, software written by Eric Thomas in 1986 (and named LISTSERV) helped automate such list-maintenance functions as adding or deleting members, and posting and distributing messages.

In its simplest form, a listserv (sometimes still called a mailing list or distribution list) enables an individual to send a message, such as announcement of a meeting, to two or more recipients. Frequently, though, postings are made by multiple members of the list, providing an electronic forum for discussion. Today, listservs are commonly used by professional organizations or groups of people sharing common interests—members of a junior soccer league, retirees who like clog dancing. Lists may be unmoderated (postings are automatically distributed without review by anyone) or moderated (someone collects messages received over a period of time and edits them in some way before posting—enumerating the topics covered, summarizing contents of the posts, or censoring objectionable material).

Listserv messages are sent and received as email. The platform tends to be associated with the workplace or with people who have workplace experience. Teenagers and college students are less likely to use listservs—or sometimes even to know what they are. Blogs and social networking sites, more often the province of youth, can be configured to accomplish roughly the same goals.

Blogs (=Web Logs)

While email took more than twenty years to reach a basic level of domestication, blogs were nearly an overnight sensation. The word *blog* comes from "web log," a term coined by Jorn Barger in 1997 to refer to a list of web-site URLs that the person creating the list found to be of interest and wished to share with others. Some web logs consisted of a set of headlines, followed by links to the original sites. Others offered brief news summaries or discussions of contemporary issues. In either event, frequent updating was common.

In short order, "web logs" morphed into blogs, and the genre exploded in popularity. Helping fuel this revolution was the introduction of software tools (often free) enabling average Internet users to create their own blogs without knowing HTML (hypertext markup language—the computer language in which much of the coding for web sites has traditionally been done). By the mid-2000s, blogging platforms such as Blogger and LiveJournal were encouraging teenage girls to keep online diaries, would-be social critics to get their political two cents in, stay-at-home mothers to share secrets for toilet training, and undergraduates studying abroad to update friends back home. Chapter 6 analyzes blogs in more detail, with a special eye toward understanding the roots of their appeal.

MySpace, Facebook, and YouTube

Among the newest arrivals on the one-to-many asynchronous communication scene have been social networking sites such as MySpace and Facebook, along with video counterparts like YouTube.[11] While some of these platforms (especially Facebook) historically restricted the community of users having access to information posted, others (such as YouTube) have always been open to the general public. Chapter 5 takes a closer look at Facebook. The YouTube phenomenon is addressed in chapter 6.

One-to-Many: Synchronous

The last category of electronically-mediated communication brings us a step back in time. For nearly twenty years (all of the 1980s and much of the 1990s as well), one-to-many synchronous communication was perhaps the best-known context for communicating online. Some of these platforms (such as chat) still exist, but others have largely been replaced by newer pursuits, including massive multiplayer online role-playing games such as World of Warcraft.

Computer Conferencing

Long before there were personal computers, even before the general public thought computers had anything to do with their lives, Murray Turoff was looking to harness computer power to enable people in dispersed physical locations to communicate in real time. The year was 1971. Turoff worked for the U.S. Office of Emergency Preparedness, which was interested in developing decision-making communication systems for use under catastrophic circumstances, including nuclear attack. Turoff devised a scheme known as EMISARI (Emergency Management Information Systems and Reference Index), which used a mainframe computer to link participants around the country who were sitting at teletype terminals (think of glorified electric typewriters) connected via long-distance telephone lines. EMISARI worked like a text-only telephone conference call.[12]

Today, text-based conferencing has largely been eclipsed by other communication tools, along with more sophisticated versions of the traditional telephone conference call. Inexpensive telephone rates, speaker phones, and video conferencing (not to mention Internet telephone protocols such as Skype) make lengthy textual exchanges in real time feel as old-fashioned as ship-to-shore radios. Computer-based technologies sometimes operate simultaneously but sub rosa. Lawyers and business people commonly shoot emails or IMs to colleagues during telephone conference calls, offering advice about what to say or topics to avoid. When the proceedings get boring, the same players send personal messages to make use of slow time or even stay awake.

MUDs (Multi-User Dungeons/Dimensions) and MOOs (MUDs, Object Oriented)

Readers who remember the Watergate break-in or when the Beatles first sang "Yesterday" may well know about MUDs. These online adventure games were originally modeled upon "Dungeons and Dragons," a fantasy role-playing game from the early 1970s. MUDs are synchronous environments in which multiple players interact within a textually created imaginary setting.

The first such online adventure game was created in the late 1970s by Roy Trubshaw and Richard Bartle at the University of Essex.

Why "MUD"? Originally, the letters stood for "Multi-User Dungeons." Over time, the acronym came to be more neutrally billed as "Multi-User Dimensions."

When the early MUDs were designed, computers had very limited graphics capability. Players were necessarily restricted to verbal descriptions of scenes, actions, and emotions. Unlike newsgroups (which talked about the real world, using asynchronous postings to the public at large), MUDs allowed a comparatively restricted circle of participants to move synchronously through scenarios of their own construction. Players assumed pseudonyms and interacted according to preestablished navigation rules for traversing a defined terrain.

For their first decade, MUDs were heavily dominated by male players engaged in otherworldly adventures. Over time, MUDs began expanding to include wider ranges of participants and more social functions. Object-oriented programming was introduced into MUDs, yielding the concept of MOOs (translation: MUDs, Object Oriented), so named in 1990 by their creator, Stephen White at the University of Waterloo. That same year Pavel Curtis extended the programming power of MOOs through a program called LambdaMOO. MOOs commonly are based on real-world locations (a university campus, a house), inviting participants to speak and act within particular zones (such as a room or a walkway). By the mid 1990s, MOOs were appearing in social and educational contexts, and graphics and sound were introduced as well.

Today, some die-hard early gamers continue to do combat in MUDs, although most have moved on to sophisticated commercial online multiplayer games or Second Life. Some educational MOOs are still around, but their heyday has ended.

Chat

If MUDs and MOOs created virtual worlds through which to move, chat was created simply to converse. Generically, chat is a synchronous platform for holding conversations with multiple participants. Early precursors included Turoff's EMISARI and then, in the 1980s, UNIX-based "talk" programs, which allowed several users to engage in instant messaging simultaneously. However, chat as we now know it wasn't born until 1988, when Jarkko Oikarinen, a student at the University of Oulu in Finland, wrote a program that came to be known as Internet Relay Chat (IRC), intended as an improvement on UNIX "talk." By the early 1990s IRC became known to the wider public, serving as a template for more generic chat programs available through Internet providers such as America Online and through the web.

Similar to the protocol for newsgroups, participants in chat enter into a "channel" (for IRC) or "room" (for AOL), ostensibly dedicated to a specific topic. With chat, however, the medium is synchronous. It also invites both playful and manipulative behavior. Users log on through nicknames (akin to participation in MUDs), free to camouflage their real-world identities, including age, gender, and personal background. While conversation takes place in real time, users can (as with newsgroups) scroll back through the archive to respond to earlier postings. Like newsgroups, listservs, blogs, and MUDs or MOOs, chat generates a quasi-public textual record.

Chat rooms became immensely popular in the United States in the 1990s, as a growing number of home-computer users paid their monthly fees to be connected to the Internet and then found themselves glued to the computer screen for an evening's entertainment. But then two things began to happen, causing chat rooms to lose some of their appeal. The news media issued troubling reports of people (commonly teenagers) being lured from online conversation with strangers into real-life encounters, sometimes with deadly results. At the same time, alternative online forums with more controlled access (such as educational MOOs or collections of buddies on IM) offered social alternatives. When I ask my American college students if they enter chat rooms these days, a typical answer is, "No, I did that as a kid, but not anymore. It's too creepy out there."

Second Life

Besides one-to-many asynchronous social networking sites such as Facebook, MySpace, and YouTube, the place to be in 2008 for synchronous virtual action is Second Life. Second Life might be thought of as a cross between a massive multiplayer online role-playing game and the DreamWorks Animation studio. Designed by Philip Rosedale's Linden Lab (originally located on Linden Street in San Francisco) and opened to the public in 2003, Second Life describes itself as a "3D online digital world imagined, created, and owned by its residents."[13]

Participants in Second Life build alter egos (as with MUDs), but they create a great deal more as well. You may purchase land (for real money, exchanged for Linden dollars), hawk goods and services (even making a decent real-world living),[14] go off on wild virtual adventures, or use a Second Life site for teaching a university course.[15] Such commercial enterprises as Sears and Circuit City are building virtual stores—for selling real products.[16] By mid-2007, Second Life boasted more than seven million members.

• • • ELECTRONICALLY-MEDIATED COMMUNICATION: A GROWTH INDUSTRY

So many ways of harnessing computers (and computer technology) to communicate. But how many people use which types?

Like trying to lash the wind, attempting to provide up-to-date usage statistics is a treacherous task. Reliable tallies are often a year or two out of date by the time they are issued. Add in the period between which I pulled these statistics and you picked up this book, and the gap starts widening into a chasm.

What to do? Statistics are useful, even if they aren't absolutely current. They establish benchmarks against which we can measure future development of electronically-mediated communication. What's more, comparing statistics across national and cultural boundaries offers insight into how and why patterns of online and mobile communication have taken different turns in diverse social groups.

Before doing the numbers, we need to prepare ourselves for methodological inequities and uncertainties. Two research organizations may ostensibly be measuring the same phenomenon, when in fact their studies don't control for identical variables. If you are counting how many Americans use text messaging, do you include people who have tried texting once or twice (then abandoning it) or just those who are regular texters? If you're tallying how many IMs are sent a month, have you included IMs sent on mobile phones rather than only via computers? We have reasonably reliable figures for some countries or demographic groups, but not trustworthy information on others.

With these statistics, caveat emptor. Take them as general indicators of some contemporary usage patterns, not as comprehensive or as gospel truth.

Internet Access

Putting aside web-enabled mobile devices, you need access to a personal computer connected to the Internet to engage in online language. According to Internet World Stats, as of March 2007, 70 percent of the American population used the Internet, compared with 39 percent of Europeans and 17 percent of the world as a whole.[17] Historically, the Internet was largely the province of English speakers (and the English language). In 1996, English was the native language of 89 percent of Internet users. By 2006, more than two-thirds of those on the Internet were native speakers of some language other than English.[18] Internet World Stats reports that 30 percent have

English as their native language, 14 percent are native speakers of Chinese, 8 percent are native speakers of Spanish, and nearly 8 percent are native Japanese speakers.[19]

Statistics on Computer-Based Communication Platforms

Once people have Internet access, in what types of online communication are they engaged? The predominant language application is still email. According to Ferris Research, six billion business emails were sent in 2006.[20] Even without knowing precisely what counts as "business," six billion is a huge number.

IM has also become a pervasive platform for one-to-one online communication. ComScore Media Metrix reported that as of May 2006, there were nearly 340 million people worldwide with instant messaging accounts.[21] Given the number of teenagers and young adults (at least in the United States) who maintain multiple IM accounts, that number could be slightly high, but even an estimated one out of every twenty people on the planet communicating by IM is formidable, especially considering that the modern medium is barely a decade old. (The world's population is roughly 6.6 billion.)

Blogs and Facebook are good examples of asynchronous one-to-many platforms whose usage has soared. While no one really knows how many blogs are out there, Technorati reported in May 2007 that it was tracking 83.1 million blogs, and that more than 175,000 new blogs were being added every day.[22] As for Facebook, comScore Media Metrix indicated there were 24 million members by spring 2007.[23]

Mobile Phone Statistics

Computer-based communication tells only part of the story of how electronically-mediated language is being created and conveyed. The other major technology is obviously the mobile phone. How many mobile phones are there?

This turns out to be a rather complicated question. Over the years, I have owned four phones in the United States and one that I purchased in the UK. So how many phones should you count me as having? My American handsets were owned seriatim, with all four connected to the same phone contract. Obviously, what we want to tally is phone subscriptions, not handsets.

But the counting problem is still not solved. The phone I bought in Brighton, England, a Nokia 1100, can be used in all countries that are on the GSM system. I just need to purchase a new SIM card (the small computer chip that fits into the back of the phone) each time I enter another country. Over the years, I have bought SIM cards in the UK, Italy, Spain, Greece, Switzerland, Australia, Sweden—and then Italy and Spain, all over again. To purchase a card, I register paperwork with the vendor, thereby creating a new subscription (with a new phone number). If I remain in the country for awhile, I may top up the card by adding money. However, the cards expire if not used within a set length of time. Because more than a year elapsed between my two visits to Italy, and between my visits to Spain, I needed to purchase entirely new SIM cards—complete with new registrations and telephone numbers.

Technically, I have had nine mobile phone subscriptions over the past four years. The fact that these subscriptions were seriatim (like my ownership of handsets in the United States) does not get reflected in the official tallies. Mercifully, most mobile phone users are less fickle than I, though by now it should be clear that statistics on mobile phone usage need to be taken with a hefty grain of salt.

The International Telecommunication Union (ITU) offers the most exhaustive statistics on mobile phone subscriptions worldwide. Because the ITU also provides historical information, along with tallies on landline phones, we can look at current mobile phone data in comparative context. The statistics below were taken from the ITU web site, accessed in June 2007.[24] In each case the data (which I have rounded) represent the number of people, per 100 inhabitants, who had either a landline or mobile phone:

	Landline Phones		Mobile Phones	
	2000	2005	2000	2005
US	68	59	39	72
Europe	33	41	37	86
World	16	19	12	34

Several observations jump out from these figures. While landline subscriptions are gradually creeping upward in other parts of the world, they are actually falling in the United States. The reason? Mobile phones are beginning to replace landlines. In the latter half of 2006, nearly 13 percent of American homes had only a mobile phone.[25] Many among those ranks were young adults. A study from early 2007 suggests that among eighteen- to twenty-four-year-olds, roughly one in four had only a mobile phone.[26]

Second, mobile phones are becoming ubiquitous. If we simply look at subscription numbers versus population statistics, more than one out of every three people on the face of the earth has a mobile phone subscription. Overall, the proportional subscriptions in Europe remain greater than in the United States. According to the ITU, in a number of countries in Europe and beyond, the calculations exceed more than 100 mobiles per 100 people. (Among the clearly multimobiled are Luxembourg: 155, Italy: 124, Hong Kong: 124, and Israel: 112.)

Why would anyone need more than one mobile phone subscription (or phone)? The reasons vary. In the Far East, employees sometimes have a "boss phone" (reserved just for communicating with one's supervisor at work), a second phone for interactions with other business associates, and yet a third for family and friends. Multiple subscriptions can be a cost-saving strategy: Use one SIM card (which means one subscription) for most calls, because it has fairly cheap rates; use another SIM (entailing another subscription) for contacting people on the same telecommunications system such as Vodafone or Telia, because these calls or text messages are free. SIM-switching is particularly common in the developing world, where cost is of major concern.

Text Messaging Traffic

Once you have a mobile phone, there are many uses for it: talking with other people, checking the weather, playing games, purchasing food from vending machines, doing banking, listening to music, and, of course, communicating through text messaging. How many text messages are being sent? Accurate statistics are not easy to come by, especially because the numbers keep multiplying so rapidly.

Here are sample estimates:

- in 2005, more than one trillion text messages were sent globally[27]
- in 2006, Americans sent 158 billion text messages, which was nearly double the number sent the previous year[28]

At the same time, Americans continue to talk on their mobile phones, in part because they pay for massive numbers of voice minutes each month, whether they use them or not. According to the CTIA (the Wireless Association in the United States), customers used 1.7 trillion minutes of talk time in 2006, up 20 percent over 2005.

• • • LANGUAGE IN THE NEW MILLENNIUM

Like typewriters and landline phones before them, computers and mobile phones convey language. But what does the language itself look like?

The earliest discussions of computer-mediated communication debated whether online communication was a new form of language—or a degenerate one. Public discussion in the late 1980s and early 1990s focused on traits such as abbreviations, acronyms, emoticons (also known as smileys), and untidy sentence mechanics (in spelling, punctuation, and grammar). There was also much talk about flaming, that is, using rude—even crude—language.

As CMC moved beyond the academic world in the 1990s, and more everyday users (from teenagers to small business people to grandparents) began using email, listservs, newsgroups, chat, and eventually instant messaging, the tenor of analysis began to shift. The question now was whether CMC in general—or at least email or IM in particular—more closely resembled speech or writing. Overwhelmingly, the verdict was "speech" (based largely upon the fact that messages tended to be informal), though the arguments were typically based on isolated examples—a smiley face here, a *btw* there—rather than upon empirical research.

There were some exceptions. As the phenomenon of online language started to attract academic researchers, solid evidence began mounting regarding the ways in which CMC shared features with formal or informal writing, and with formal or informal speech. Susan Herring's 1996 landmark collection, *Computer-Mediated Communication: Linguistic, Social, and Cross-Cultural Perspectives*, set the standard for the decade of research that followed.

With the continued expansion of email and IM in the late 1990s, and (in the United States) the gradual introduction of text messaging in the new millennium, public discussion began to shift from the linguistic nature of electronically-mediated communication to the effects this sort of language might be having on everyday offline writing. Were IM and texting, especially as practiced by teenagers and young adults, ruining their ability to craft respectable school essays? Were traditional standards of spelling and punctuation (not to mention logical coherence) soon to be left in the dustbin of history?

In February 2005, I chaired a symposium on "Language on the Internet" at the annual meetings of American Association for the Advancement of Science. In preparation for our session, I asked a distinguished panel of experts to think about concrete ways in which language conveyed via online and mobile devices might be influencing spoken and written language. Our consensus was that beyond a few acronyms such as *brb* ('be right back') surfacing in some people's speech, or *lol* ('laughing out loud') or smiley faces popping up in more informal offline writing, the actual linguistic impact of electronically-mediated communication was surprisingly small.

If online and mobile language are not having sweeping effects upon everyday language, then what is the big deal? Are IM, blogs, text messaging, and the like simply interesting curiosities? Or is there more to the story?

In this book I argue that these new forms of language are having profound impacts upon both the linguistic and social dimensions of human interaction. I begin making the case in the next chapter, which lays out how online and mobile language is empowering and emboldening us in the ways we control the "volume" on our communication with others.

3

• • • Controlling the Volume

Everyone a Language Czar

He looked like an intelligent fellow, wrinkled shirt and unkempt hair not-withstanding. A college freshman, he was part of a focus group I was observing on how students use the features of instant messaging. When the group facilitator got to the part of the session dealing with parents and IM, a collective howl went up among the half-dozen participants. Most parents, they said, failed to understand that IM is not the same thing as email. The students groaned over how long their moms and dads took to compose messages, unaware of the staccato-paced missives their progeny typically wrote, and equally oblivious to how much multitasking their sons and daughters engaged in while waiting for their parents to finally press "enter."

Then the conversation turned to away messages (those presumably brief announcements many IM users post for other members of their buddy list, ostensibly signaling a temporary absence from the computer). The tousled student claimed the floor: "The worst thing I ever did was teach my mom to IM over Christmas. Now she's IMing me all the time, wanting to know where I am and what I'm doing."

His compatriots nodded sagely. After all, these are the same people who hear their mobile phones ring, glance at the number on the display screen, and then ignore certain calls, nonchalantly announcing, "It's only my mom." But he went on: "The most annoying part is that she still thinks I'm, you know, her little innocent kid. Sometimes I want to post away messages that, well, I don't want her to read. [Translation: "containing profanity and/or alluding to behaviors illegal for someone under twenty-one."] So I just block her."

Again, signs of group accord, and then a few more details: "Once I've had a message up for awhile and I know my friends have seen it, then I take it down and unblock her. She's like, 'Where have you been? I've been trying to reach you for hours.' And I go, 'I've been busy. You know, I'm a student,' and eventually she drops it." As we've explained, when you block a member of your IM buddy list, you appear (electronically) to that person as if you aren't signed on and therefore can't receive an IM.

College students are not the only ones engaging in discourse management. For many adult professionals, that management is increasingly unidirectional: always on. The spirit of needing to be constantly accessible—and accessing messages—is epitomized by the slogan of RIM (Research in Motion), makers of the BlackBerry: "Always on, always connected."

An apt metaphor for analyzing technologies and techniques for manipulating conversation is the volume control on a radio or television. In electronically-mediated communication, users turn up the "volume" when they incessantly check their email. Alternatively, we can turn down the "volume," as do many of my students in choosing to ignore an incoming call on their mobile phones or precluding a potential communiqué from Mom by blocking her on IM. In talking about communication, the volume control image refers less to physical noise level than to frequency of contact or restrictions on access.

Along with the volume-control metaphor, another useful concept is affordances, a notion originally developed by the psychologist James Gibson and later applied to technology issues.[1] Affordances are the physical properties of objects that enable us to use them in particular ways. For instance, an affordance of paper—unlike computers—is that you can record writing on it without needing electricity. An affordance of mobile phones is that they allow us to roam around, not tethered to a landline connected through actual telephone wires.

One way to magnify the affordances of language technologies is to multitask. Linguistic multitasking takes many forms—from simultaneously talking on the phone and reading email to participating in multiple IM conversations. Multitasking is widespread in contemporary society, especially when using computers. Our question in this chapter is how multitasking, while using language technologies, enhances our ability to orchestrate interpersonal communication.

We begin by looking at discourse control in traditional face-to-face and written-communication settings, and then at the affordances new language technologies add for adjusting the conversational "volume." Finally, we turn to the cognitive and social aspects of multitasking, specifically with respect to IM and mobile phones.

• • • MANNING THE CONTROLS

Although speech and writing are social activities, people aren't inherently continually connected with one another—that is, they aren't "always on."

Individuals have always developed strategies for controlling their interactions with other people, including establishing zones of privacy, even in societies offering little physical space for seclusion.[2] What shifted over time are the amount of control and the mechanisms for effecting it, reflecting both new technologies and the ways they let you multitask.

Language users manage their communication in three sorts of ways. The first involves access: increasing our chances of actually talking with particular people. Another is avoidance mechanisms for averting linguistic encounters. And the third is manipulation, such as putting your boss (who is in the process of violating company policy) on speaker phone. All these maneuvers have a role in adjusting the "volume" on spoken or written language.

Speakers and listeners have historically been at the mercy of the laws of physics and the social pecking order. As for access to others, the human voice only projects so far, even with cupped hands or megaphone. To avoid talking to some people, those in positions of authority could typically restrict who got to speak with them, while the rest of the public was more exposed to unregulated encounters—on the street, at the marketplace, in church. In response, people have devised social avoidance mechanisms: crossing the road or looking in shop windows when attempting to avoid conversation with someone heading their way, or offering a brief greeting before dashing off to made-up engagements.

"It's too late, Roger—they've seen us."

Conversely, we sometimes take advantage of social conditions to become privy to the conversations of others. Eavesdropping is an age-old practice—whether by intention or accident.

What about "volume control" for written language? Traditionally, access was limited by physical or economic circumstances. Ships carrying the mail sometimes sank; roads on which mail coaches traveled were filled with brigands; postal rates were high.[3] At the same time, both letter-writers and recipients exercised their own access control. Senders paid for mailing options such as express delivery or "return receipt" to speed transmission or increase the chances of getting the recipient's attention. Those on the receiving end might delay responding or ignore the missive outright.

Written letters or memoranda enable people to avoid face-to-face encounters. From classic "Dear John" letters breaking off romantic relationships to impersonal job termination notices, writing provides a social shield, enabling us to avoid delivering unwelcome news in person.

Finally, writing generates opportunities for deception or gossip. Rather than accurately depicting themselves, correspondents sometimes misrepresent their physical appearance or academic credentials. Instead of maintaining the presumed confidentiality of a letter addressed to a specific person, recipients have long shared documents with others for whom the writing was not intended.

Opportunities for control slowly multiplied with the introduction of technologies that bridge physical distance. The operative word here is "slowly." Start with landline phones. Alexander Graham Bell's original telephone, patented in 1876, provided little control to those receiving phone calls. Since the instrument had no ringer, the telephone connection was always "on." Even after ringers were introduced in 1878, those with telephones remained at the social mercy of callers, since the phone needed to be answered.

By the late nineteenth century, telephone connections in houses and workplaces of the well-to-do were mediated by servants and secretaries. Yet even as phone access became increasingly affordable and home subscriptions rose in the twentieth century, a ringing phone remained a summons that nearly always took precedence over an ongoing face-to-face conversation. Those with telephones were "always on" and potentially "always connected." This situation only eased with the invention of the answering machine. While prototypes existed as early as 1900, not until 1971 was there an inexpensive model geared to the consumer market.[4]

Over the last thirty years, technological developments have provided phone users with increased opportunities for controlling conversation. With the development of answering machines (rechristened "voicemail"), we became able to screen incoming calls and to leave messages for people who weren't home. Through the Internet, we procure direct telephone numbers, enabling us to bypass traditional intermediaries such as secretaries. Using call-waiting features, we line up for someone's attention.

Telephone technologies also decrease access. People with whom we wish to speak can choose to avoid contact through caller ID or by blocking calls from specific numbers. Call initiators avoid conversations by using express messaging, whereby a connection goes directly to voicemail without the recipient's phone ever ringing. In the business and professional worlds, telephone systems often preclude our speaking with a human being, shunting us instead to phone trees, voice recognition systems, and recorded messages.

Besides opportunities for increasing or decreasing access, modern phones also provide tools for manipulating communication from other people. Leave me a nasty voicemail, and I might forward it to your worst enemy.

Computer-based language technologies build upon the volume-control options of modern landlines. For example. we use the Internet to locate electronic addresses of strangers. Anecdotal evidence suggests many people are more likely to reply to email from an unknown correspondent than to an unsolicited letter or phone call. Why? It takes less effort to answer someone online. What's more, the social distance afforded by email makes responding a less personal act than a face-to-face or even voice-to-voice encounter.

Email is a handy volume-control tool in other ways as well. Again, as with modern landline phones, email allows us to avoid or manipulate communication we've received. We leave incoming messages unread (or unanswered) for as long as we please or local social conventions permit. Then there is manipulation: Email can be forwarded—more potential for gossip—to unintended recipients, reminiscent of callers unknowingly being placed on speaker phone.

Instant messaging introduces a new set of control mechanisms. Even if you know a person's IM screen name, people can make themselves appear to be offline by blocking you. At one remove, you can keep in touch with friends by reading their profiles or away messages rather than directly contacting them through an IM or phone call.[5] Similarly, users wanting to avoid face-to-face or even telephone voice contact with acquaintances can IM them instead.[6]

Finally, what about controlling conversation via mobile phones? As with the BlackBerry, whose users tend to be "always on," it's sometimes not clear who is controlling whom with mobile phones, which enable people to be in "perpetual contact" with one another.[7] However, having your mobile phone turned on need not imply you welcome being generally available. What American college students seem to like most about mobile phones is being able to access others, while what they like least is other people always being able to contact them.[8]

As devices for spoken language, mobile phones afford a variety of control features transcending those available on landlines (such as caller ID, call waiting, or speaker phone). Owners of mobile phones can assign distinct ring tones to each person in their address book, making it unnecessary even to view the phone display panel before deciding whether to take a call. What's more, camouflage services are available that, for instance, provide background noise from a traffic jam, enabling someone to say with authority, "Sorry, I'll be two hours late. I'm stuck in traffic" (while actually sitting at a café). These services can also generate a ring tone in the middle of a conversation (appearing to signal an incoming call), providing a plausible excuse for terminating the current exchange.

Text messaging on mobile phones introduces another set of control mechanisms. The access and avoidance issues with texting are similar to those with email and IM in that users can identify the message sender before deciding how and when to respond. Sending a text message rather than placing a voice call is often done to eliminate small talk and save time (see chapter 7).

• • • MULTITASKING

We've been talking about a variety of ways in which people choreograph their spoken and written communication with one another, increasingly with the aid of technology. But there is another crucial tool we have for manipulating language give and take, and that is to do something else at the same time—to

"Are you multitasking me?"

multitask. Because those on the receiving end of emails, IMs, or phone calls can't see us (webcam technology excluded), they often are unaware when we engage in additional activities. To rephrase Peter Steiner's famous 1993 *New Yorker* cartoon ("On the Internet, nobody knows you're a dog"), we might argue that "On the phone and on the Internet, nobody knows you're multitasking." OK, sometimes they do figure it out.

Multitasking—making simultaneous demands upon our cognitive or physical faculties—is a common enough necessity in everyday life. Think about what's involved in driving a car. We need to look three ways (ahead, in the rearview mirror, and peripherally), while controlling the speed and direction of the vehicle, and perhaps conversing or listening to the radio or a CD. For seasoned drivers, operating a vehicle may not seem like a skill requiring multitasking, but try teaching someone to drive, and the complexity of doing several tasks at once unfolds before your sometimes terrified eyes. ("How can you expect me to see the car on my right? You said to look ahead!") A less life-endangering example is playing the piano or the organ. The musician needs to read multiple lines of musical notation and control two hands, along with one or two feet.

Another reason for multitasking is our perception of demands on our time. Time-driven multitasking (such as house cleaning plus child care, or commuting on a bus plus reading) is everywhere.[9] Time-interval diary studies in the UK suggest that through multitasking, people "add" nearly seven hours of activity to each day.[10]

Sometimes multitasking is a response to an emotional state such as loneliness. Many people turn on a radio, music player, or television upon returning home or entering a hotel room, even though their primary activity is neither listening nor viewing. We'll also see that impatience and boredom are common motivations to multitask while using language technologies.

Multitasking can involve either mental or social activity. Sometimes we perform two or more cognitive tasks at once, such as doing a crossword puzzle while completing a questionnaire. Other times, we participate in more than one interpersonal activity, say alternating between a face-to-face conversation and typing an IM. And of course, we can multitask by combining mental tasks (like doing homework) with social activity (maybe a phone conversation). To keep our terminology straight, we'll speak of "cognitive multitasking" when we are looking at the mental consequences of doing more than one thing at the same time. We'll use the term "social multitasking" when we're interested in the social effects of doing two (or more) things at once.

Psychologists have been interested in cognitive multitasking for decades.[11] Most studies have suggested that engaging in simultaneous activities (particularly involving unfamiliar or unpracticed tasks) decreases performance level. For instance, watching television while simultaneously recalling sets of digits or while doing homework makes for poorer recall (and homework results) than focusing on a single task.[12] Similarly, switching between tasks (such as alternating between solving mathematics problems and classifying geometric objects) degrades performance.[13]

Laboratory tests indicate that if the multiple tasks tap different modalities (say, one involving seeing and the other involving hearing), degradation in performance is generally less than when both tasks rely upon the same modality (for instance, both visual).[14] In the real world, other factors may come into play, such as people's experience in processing particular multiple stimuli. Take listening to music while studying. Students who typically study with background music are, under test conditions, more successful at learning material when music is playing than are students who study where it is quiet.[15] And not surprisingly, the nature and difficulty of the tasks at hand sometimes temper the effects of cognitive multitasking.

Multitasking is becoming increasingly common among adolescents and young adults, especially when it comes to multitasking involving media such as computers, video, and music (along with that old standby, television). In 2005, the Kaiser Family Foundation reported that American children aged eight to eighteen used media for recreational purposes (that is, not school-related) an average of almost six-and-a-half hours a day. When you factor in multitasking (for instance, using a computer while watching television), these children reported being exposed to media for more than eight hours daily.[16]

The Kaiser report gives some eye-opening statistics on just how much multitasking children do: "Depending on the medium about which they were asked, from one-quarter to one-third of adolescents report using multiple media 'most of the time.' " If we combine the categories "most of the time" and "some of the time," we find that seventh- through twelfth-graders multitask when they are using a computer (62 percent of subjects), listening to music (63 percent), watching television (53 percent), and reading (58 percent).[17]

With the proliferation of communication technologies such as computers and ever-ringing landline and mobile phones, researchers have begun amassing data on the negative impact of modern cognitive multitasking. Even when people attempt to attend strictly to a single task such as working on a report, they are often distracted by other cognitive demands—with dramatic results. Psychologist Glenn Wilson administered a variety of tasks, including IQ tests, to ninety subjects in the UK. When these tasks were performed in the presence of communication distractions such as a ringing telephone, average performance on the IQ test fell ten points—essentially the equivalent of missing an entire night's sleep.[18]

Several investigations have explored the relationship between academic performance and use of the Internet. One study reported degraded memory for lecture content when students simultaneously listened to classroom lectures and accessed the Internet to do searches or communicate with colleagues online.[19] Another study found that undergraduate students with Internet access in their dormitory rooms engaged in considerable multitasking. What's more, the ratio of school work to recreational computer-based activity was roughly one to four—hardly an efficient way to complete assignments.[20]

Another cluster of experiments has explored the cognitive effects of interrupting a person's work flow when engaged in a computer-based activity. What happens, for example, if I send you an IM right when you are trying to do an online search? Research suggests that the timing and form of these kinds of interruptions are critical in determining how disruptive the incoming message turns out to be.[21]

Is multitasking with communication technologies necessarily detrimental to cognitive performance? The answer may reflect the extent to which people think of themselves as doing multitasking. One of my former students argues that it makes little sense to talk about multitasking on a computer. In his eyes, computers are naturally multitasking devices. (By analogy, recall how driving an automobile or playing the organ puts simultaneous demands on our cognitive and physical faculties.) Having grown up with the technology, he doesn't perceive a degradation of performance by engaging in simultaneous computer-based activities (such as surfing the web and writing a paper or carrying on an IM conversation). Drawing upon the notion of domestication, computers are domesticated technologies for much of his generation—though whether computer multitasking actually degrades performance within this age cohort remains an unexplored question.

Putting aside teenagers and young adults, what about the rest of us? In learning to drive a car, the ability to look three places at once develops with experience. A growing literature documents how practicing complex skills, such as learning to juggle balls, leads to changes in adult brains.[22] My favorite case, though, is London taxi drivers.[23]

London is renowned not only for its fleet of black taxis but for the ability their drivers have to locate addresses in the city. This ability is no accident. Would-be drivers sometimes train for several years before being certified as having what is known as "the knowledge." A team of researchers at University College London was curious to know if such training—and subsequent experience on the job—resulted in measurable changes to cabbies' brains. The answer turned out to be yes. The posterior hypocampus was found to be larger in cab drivers than in control subjects. Drivers who had been in the business for forty years showed even larger brain areas than those new to the profession. By analogy, perhaps if we practice multitasking with language technologies, our brains will adapt, and performance on all the cognitive tasks will be laudatory.

But what about social multitasking? Does it degrade social performance? Think of talking on the telephone while conducting a web search or doing IM. Does the quality of the IM conversation or the spoken exchange suffer? Unlike the case of cognitive multitasking, there is little research on the interpersonal effects of multitasking while communicating with others. What is clear, though, is that social multitasking involves volume control over the communication, such as mumbling "uh huh" to the person with whom you're speaking so you can focus on making an online purchase or on deciding which of three IM messages to respond to first.

• • • MULTITASKING AS VOLUME CONTROL

Data on contemporary multitasking behavior involving language technologies are beginning to trickle in. A few studies (including research done by Nancy Baym and her colleagues, and by the Pew Internet & American Life Project) have asked participants to note their multitasking activities while using computer-mediated communication.[24] Retrospective self-reported behavior is, however, notoriously inaccurate. To help address this methodological challenge as well as to gather data specifically involving instant messaging, my students and I undertook pilot studies on multitasking while communicating via IM.

Using an online questionnaire, we charted the multitasking activities of American University undergraduates who were engaged in IM conversations. The data were collected in fall 2004 and spring 2005. We knew that all subjects were participating in at least one IM conversation when they completed the questionnaire, since IM was the medium through which student experimenters distributed the URL for the questionnaire web site. (Subjects were members of the experimenters' buddy lists.) Most participants were in their dormitory rooms at the time they received the questionnaire, affording ample opportunities for involvement in multiple activities.

The results revealed a high level of multitasking. In our first study, out of 158 subjects (half male, half female), 98 percent were engaged in at least one other computer-based or offline behavior while IMing:

Computer-Based Activities

Web-based activities:	70%
Computer-based media player:	48%
Word processing:	39%

Offline Activities

Face-to-face conversation:	41%
Eating or drinking:	37%
Watching television:	29%
Talking on the telephone:	22%

Subjects often participated in multiple examples of the same activity (such as having three web applications open or being involved in more than one IM conversation). Students in this first study averaged 2.7 simultaneous IM conversations, with a range from 1 to 12.

Common sense dictates that people can't literally participate in multiple IM conversations simultaneously. And indeed they do not. Subsequent focus groups revealed that many of the students used IM both synchronously and asynchronously, that is, turning the volume up or down on particular conversations. Decisions depended upon such factors as "how good the gossip is" in a conversation, how serious the conversation is, and individual communication habits. A few students found it rude to hold simultaneous IM conversations, though they were by far in the minority.

We used both informal focus groups and a revised online questionnaire (this time with fifty-one subjects) to probe why students multitask while using a computer. Most respondents spoke of time pressures: Multitasking enabled them to accomplish several activities simultaneously. Shortage of time was also invoked to justify concurrent IM conversations. Interestingly, several students commented that IM is not, by nature, a stand-alone activity. When asked whether they ever held a single IM conversation during which time they did not engage in any other online or offline activity, the overwhelming response was no. Such behavior, said one participant, would be "too weird," because IM conversations are, she continued, conducted as background activity to other endeavors. A number of students from the second online study noted they multitasked on computers because the technology enabled them to do so. As one woman put it, "There is no reason not to when everything is accessible at once."

Ten of the fifty-one students in the second online study indicated that they multitasked while using computers because they were bored. Boredom sometimes resulted from having to wait for the person with whom they were

IMing to respond. Other students spoke of "get[ing] bored with just one activity" or "having too short an attention span to only do one thing at a time."

Focus-group members observed that with IM, students are in control of how dynamic a given IM conversation is. With lengthy IM dialogues, users may go through spurts of communication interlaced with periods of inactivity. One student aptly described IM as "language under the radar," meaning it resides in the background of other online or offline endeavors. Users control whether to make a particular conversation active (synchronous) or let it lie dormant (asynchronous), without formally closing the interchange.

We asked a series of free-response questions regarding multitasking behaviors that the students felt were or were not suitable. A typical response to the question "For which computer-based activities is multitasking appropriate? Why?" was "IMing, listening to music, browsing the web. Those are all things that do not interfere with one another." A content analysis revealed that 86 percent of the fifty students responding to this question specifically mentioned IM or email—both forms of interpersonal communication—or indicated that any type of multitasking behavior is acceptable.

Another free-response question asked, "For which non-computer activities is multitasking not appropriate? Why?" Of the forty-four students responding, 59 percent singled out face-to-face or telephone conversations as inappropriate contexts for multitasking. This number stands in stark contrast to the 86 percent who felt that conducting an IM or email conversation while using the computer for other functions was fine.

Students offered various explanations for avoiding multitasking while speaking face-to-face or by phone. The most prevalent answer was that such behavior was simply wrong. One male student said, "It's rude not to give your full attention to someone face to face," while a female observed that "talking on the phone and [simultaneously] talking to people on the computer [i.e., IM] isn't appropriate because the person on the other phone line usually feels left out or unattended to." Since face-to-face (or telephone) conversation and IM are, in principle, both synchronous activities, successfully attending to the two tasks simultaneously can be problematic. One or another of the people on the receiving end might catch you out.

Similar feelings of personal abandonment were reported in a study conducted by Sprint in 2004. Half of the respondents said they felt unimportant when a friend or colleague interrupted a face-to-face conversation with them to answer a mobile phone.[25] The following year, Hewlett-Packard reported that almost 90 percent of office workers judged that colleagues who responded to emails or text messages during a face-to-face meeting were being rude. However, one out of three of the same respondents indicated that such behavior was both acceptable and an efficient use of time.[26] The discrepancy between activities that employees took to be rude when practiced by others and the behaviors the subjects themselves engaged in suggests that people have yet to resolve the conflicting demands of social etiquette and work pressure.

Some respondents in our IM study said multitasking was only precluded if the topic of a face-to-face or telephone conversation was clearly significant. Of the twenty-six students who were against multitasking while face-to-face or on the phone, six disapproved of the behavior only if the conversation was particularly serious or important.

Other explanations for avoiding multitasking while face-to-face or on the phone were strictly pragmatic. As one male remarked, "people [on the phone with you] get pissy about hearing a keyboard clicking." Another said, "You should devote attention to someone who can see what you are doing." This second response came from a student who believed that, by contrast, reading while talking on the phone, or cooking while on the phone, was an appropriate type of multitasking because "If they don't know, it won't hurt them."

Four students (all female) eschewed multitasking while talking face-to-face or on the phone because they weren't good at it. Another complained she was disturbed when people with whom she was speaking were doing offline multitasking: "Talking on the phone—I [don't] want to listen to someone else's TV while I'm having a conversation with them. Nor do I want to hear their music. It is distracting."

To what degree do college undergraduates actually multitask with computer-based language technologies while engaging in face-to-face or telephone conversations? Researchers at the University of Kansas found that 74 percent of their nearly 500 subjects reported multitasking with a computer while in face-to-face conversation.[27] Of the 158 students in our initial multitasking study, 41 percent were engaged in at least one computer activity while talking face-to-face, and 22 percent were simultaneously on the computer and on the phone. Clearly, many American college students control the volume on their face-to-face and telephone conversations by multitasking on computers.

• • • EVERYONE A LANGUAGE CZAR

Since the demise of Nicholas II, the word *czar* had added new meanings. In many instances, it now refers to an appointed or elected figure charged with a publicly important task. During the administration of George H. W. Bush, a Drug Czar was appointed, responsible for curtailing drug trafficking and use.

At the University of Chicago, dormitories have kitchen czars who dragoon others to keep the shared cooking space clean and Netflix czars to keep students supplied with weekend entertainment.

This modern concept of czar—the person with ultimate responsibility for and control over a social practice—is also applicable to the ways in which we orchestrate our language use today. In the age of the Internet, children who feel they are not heard in their physical social communities move to MySpace. With email, I put off responding to a colleague's invitation, while I might have felt pressured to give a snap reply on the phone. Tech-savvy students deftly compose text messages with their mobile phones under the desk, all the while smiling attentively at the instructor in the front of the classroom.

Thanks in large part to new language technologies, each one of us is becoming a language czar, with growing control over our conversational realm. In the next four chapters, we probe some of the concrete ways in which computers and mobile phones provide platforms for constructing and receiving messages. In chapter 4, the theme is IM conversations and the folk wisdom that IM is a speechlike medium. Chapter 5 focuses on presentation of self to IM buddies through away messages, and to Friends (and Friends of Friends) on Facebook. Even greater communication control is possible through blogs, the focus of chapter 6. And in chapter 7, we look at America's emerging conventions for mobile phone use.

Czar Nicholas II met an unhappy end. Are there social repercussions to our growing individual abilities to control linguistic interaction? We grapple with this question in our final chapter.

4

• • •   Are Instant Messages Speech?

The World of IM

A gaggle of sixth-grade girls spilled out of the school's front doors into the afternoon sunlight. As they scanned the carpool line, looking for the parent, housekeeper, or friend's mom who would be picking them up, they hurriedly said their daily goodbyes. One girl spotted her target, and as she tugged open the passenger-side door, she turned her head to shout to her classmates, "See you on AOL!" Because I was driving the car behind, awaiting my own child, I had an excellent vantage point for eavesdropping.

The language used here is significant. Although the girl was referring to instant messaging using America Online, the word she used was *see*. In 1999 (when the incident occurred), it was unlikely that her home computer (or those of her classmates) would have had a webcam. She and her friends would not literally be seeing each other. In fact, they wouldn't actually be talking either, despite the fact the standard way of referring to instant messaging interchanges is to "talk on IM" or to have an "IM conversation."

With the explosive growth of instant messaging in the late 1990s and early 2000s, America's teenagers and college students shifted their computer allegiances (if they had any yet) from email to IM. The new medium was synchronous, an important trait back in the days when (asynchronous) email could be annoyingly slow. IM also came with its own set of alluring features. You could choose your own screen name, create a personal profile, assemble buddy lists, and "talk" with several people at roughly the same time. What's more, you could buy into an online lingo that parents probably didn't understand, such as *pos* for 'parent over shoulder'—a warning to your friend on the other end of the connection that Mom or Dad was looming in the background, and the ongoing conversation was best put on hold.

The popular media were quick to accentuate the novelty of IM. There was story after story about the arcane abbreviations, acronyms, and emoticons peppering instant messages, with the implication that here was a whole new linguistic code that only the young seemed to know. The general assumption was that outside of those funny truncated expressions such as *ttyl* ('talk to you later') or *omg* ('oh my god'), the flow of messages was very speechlike. Cementing the popular image was the fact that IM users themselves tended to describe IM as a written version of casual speech.

But were they right? The answer is important, not as an academic curiosity but because of the dominant role IM has been playing in young people's lives. If it really has the characteristics of informal speech, then IM has the potential to chip away at the prescriptive standards of traditional written language. Alternatively, if it turns out that IM embodies relevant traits of more formal written language, then we need not be so quick to panic that the medium is sending writing conventions to the dogs.

When I began my research, a few American studies of IM had appeared or were circulating in manuscript.[1] These generally looked at social issues, such as who uses it, how often, and for what purposes. With the exception of Gloria Jacobs's research, which looked at a small number of high school girls' IM conversations, the only work I knew of examining the linguistic guts of IM was done by Ylva Hård af Segerstad at Göteborg University in Sweden.[2] Unfortunately for me, her IM platform, called WebWho, was essentially a presence-indicator for users working in a Swedish university computer lab, hardly comparable to IM systems found in American homes and college dorm rooms.

I set about to explore how American college students craft their IM conversations, with a specific interest in the speech-versus-writing question. As it turned out, another variable proved crucial as well: gender.

• • • SPEECH VERSUS WRITING

Love and marriage. Yin and yang. We pair together so many noun couplets, but the relationship between the two members is not always transparent. This truism applies equally well to speech and writing. Is writing simply a transcription of speech? No. Is writing always formal and speech necessarily informal? Obviously not. Are there conventional distinctions between speech and writing that most people can agree upon? Yes, as long as we also acknowledge that differences between speech and writing lie along a continuum rather than being absolutes.[3] We write casual personal notes and deliver eloquent orations. But conventionally, writing tends to be more formal and speech more informal. In school, for instance, we are taught not to use contractions (*let's* instead of *let us*) in writing, even though they are extremely common in speech and increasingly found in contemporary written publications.

Here are some of the main differences between speaking and writing:[4]

	Speech	*Writing*
STRUCTURAL PROPERTIES		
• number of participants	dialogue	monologue
• durability	ephemeral (real-time)	durable (time-independent)
• level of specificity	more vague	more precise
• structural accoutrements	prosodic and kinesic cues	document formatting
SENTENCE CHARACTERISTICS		
• sentence length	shorter units of expression	longer units of expression
• one-word sentences	very common	very few
• initial coordinate conjunctions	frequent	generally avoided
• structural complexity	simpler	more complex
• verb tense	present tense	varied (esp. past and future)
VOCABULARY CHARACTERISTICS		
• use of contractions	common	less common
• abbreviations, acronyms	infrequent	common
• scope of vocabulary	more concrete	more abstract
	more colloquial	more literary
	narrower lexical choices	wider lexical choices
	more slang and obscenity	less slang or obscenity
• pronouns	many 1st and 2nd person	fewer 1st or 2nd person (except in letters)
• deictics (e.g., *here, now*)	use (since have situational context)	avoid (since have no situational context)

Is computer-mediated communication a form of writing or speech? Since the early 1990s, a number of linguists have explored this question. About ten years ago, I surveyed the relevant literature on email, bulletin boards, and computer conferencing, concluding that as of the late 1990s, CMC was essentially a mixed modality.[5] It resembled speech in that it was largely unedited; it contained many first- and second-person pronouns; it commonly used present tense and contractions; it was generally informal; and CMC language could be rude or even obscene. At the same time, CMC looked like writing in that the medium was durable, and participants commonly used a wide range of vocabulary choices and complex syntax.[6]

A few years later, in his book *Language and the Internet*, David Crystal investigated many types of CMC, including the web, email, chat, and virtual worlds such as MUDs and MOOs. He compared these platforms against his own analysis of spoken versus written language. Coining the term "Netspeak" to refer to language used in CMC as a whole, Crystal concluded that "Netspeak has far more properties linking it to writing than to speech. . . . Netspeak is better seen as written language which has been pulled some way in the direction of speech than as spoken language which has been written down."[7]

My earlier conclusions and then Crystal's were based upon data drawn from other researchers, not our own empirical studies. Neither of us considered instant messaging, since at the time we did our research, no one had collected and analyzed IM samples. It was time to address this deficit.

Speech as Discourse: Introducing Intonation Units

First, a word on terminology: In conversational analysis, a "turn" is the language a speaker uses while he or she holds the floor before ceding it or being interrupted. That turn may consist of one sentence, many sentences, or just a sentence fragment, such as "Hmm."

When we talk, generally there are at least two people sharing the conversational floor. To figure out what it means to "talk on IM," we need to consider give-and-take between speakers in old-fashioned face-to-face (or telephone) encounters. Several obvious questions arise: What is the length of the conversation—in words, in turns, and in time on the clock? How long does each participant hold the floor—again, in words, number of turns, or time? The idea of holding the conversational floor generates more questions: How do people know when it is their turn to talk? What happens when two people try to speak at the same time? How do we open and close conversations? How many turns (and how much time on the clock) does it take to say goodbye?[8]

There's also the issue of whether (and, if so how) people divide their turns into smaller units.[9] Within a single turn (that is, while someone continues to hold the floor), a speaker might utter a sequence of smaller chunks, such as

chunk 1:   I was wondering
chunk 2:   whether you're coming to dinner tonight
chunk 3:   or you need to work.

Wallace Chafe refers to these spoken chunks as intonation units.[10] The primary linguistic indicators demarcating a spoken intonation unit are:

* a rising or falling pitch at the end of a clause (that is, a string of language having a subject and a predicate)
* a brief pause at the beginning of an intonation unit
* a conjunction (typically *and*, though alternatively *but, or,* or *so*) at the beginning of an intonation unit

Grammatically, the intonation unit is likely to be a clause, though some clauses extend over several intonation units.

What's the connection between spoken intonation units and IM? It turns out that in IM conversations, participants frequently break their written messages into chunks. A student in my IM study sent the following message but broke it up into five distinct transmissions, each one sent immediately following its predecessor:

transmission 1:   that must feel nice
transmission 2:   to be in love
transmission 3:   in the spring
transmission 4:   with birds chirping
transmission 5:   and frogs leaping

With both speech chunks and sequential IM transmissions, a single sentence may be constructed out of a series of pieces. Our question then becomes: Are sequences of IM transmissions analogous to sequential intonation units in spoken face-to-face conversation? If so, the analogy would support the argument that IM tends to be a speechlike form of communication.[11]

• • • GENDER AND LANGUAGE

Speech versus writing is one yardstick against which we want to measure IM. A second is gender.

The topic of gender differences in language has a long history.[12] Most studies have looked at spoken language, though a small body of research has considered evidence of gender influencing written style. Internet researchers have also begun exploring gender-based correlates of online behavior. Nearly all of this work has drawn upon one-to-many data sources such as chat, listservs, or computer conferencing. With a few exceptions,[13] we know very little about gender differences in one-to-one CMC such as email and IM.

How does gender affect language? At the most basic level, languages may restrict particular words, sounds, or grammatical patterns to males or females. In Japanese, for example, only males are supposed to refer to themselves using the first-person pronoun *boku*. Sometimes a whole language is reserved for one gender, as in Australia, where Walpiri women use a sign language that males are forbidden to learn.[14] Other gender differences result from subtle acculturation. For instance, females are commonly described as using more politeness indicators than males, while men more frequently interrupt women than vice versa.[15] A number of these differences have been documented cross-culturally.[16]

Speech and Gender

Other gender distinctions are more functional. Many linguists have argued that women tend to use conversation predominantly to facilitate social interaction, while males are more prone to converse in order to convey information.[17] In Janet Holmes's words, whereas women "use language to establish, nurture and develop personal relationships," men's use of conversation is more typically "a means to an end."[18] Women are more likely to use affective markers (such as "I know how you feel"), diminutives ("little bitty insect"), hedge words (*perhaps, sort of*), politeness markers ("I hate to bother you"), and tag questions ("We're leaving at 8:00 p.m., aren't we?") than men. By contrast, men more commonly use referential language ("The stock market took a nosedive today") and profanity, and employ fewer first-person pronouns than women.

Another aspect of speech that often breaks along gender lines is adherence to normative language standards. On average, women's speech reflects standard pronunciation, vocabulary, and grammar more than men's does.[19] A variety of explanations have been offered for these gender discrepancies. One is that women are simply socialized to speak more "correctly." Another is that because women do the majority of the child-rearing, they need to model standard language for their progeny. And in the West, where women's professional choices were historically circumscribed, positions that were broadly open to women in the twentieth century (such as teacher, secretary, or airline stewardess) required their incumbents to be well-spoken.

Writing and Gender

A handful of studies have looked at gender differences in written language. Some have been historically focused, while others have analyzed modern writing.

Douglas Biber and his colleagues have studied the relationship between speech and writing by analyzing large collections of spoken and written data.[20] One of Biber's measures is what he calls "involved" (as opposed to "informational"). This metric includes use of present-tense verbs, first- and second-person pronouns, contractions, and so-called private verbs such as *think* or *feel*. Nearly all of these characteristics are associated with speech rather than writing. The distinction between "involved" and "informational" roughly parallels the "social" versus "informative" dichotomy we have already talked about for speech.

Biber's group examined an historically varied collection of personal letters written (in English) by men and women. In both the seventeenth and the twentieth centuries, personal letters composed by women (whether to other women or to men) showed a higher index of "involved" language than did letters written by men (whether to women or to other men).[21]

Using more contemporary data, Anthony Mulac and Torborg Louisa Lundell studied impromptu descriptive essays written by college students. The assignments were coded with respect to seventeen linguistic features, including "male language variables" (such as judgmental adjectives, elliptical sentences, and sentence-initial conjunctions or filler words) versus "female language variables" (for example, references to emotion, sentence-initial adverbials, uncertainty verbs, or hedge words). The investigators found that by analyzing the language used in the essays, they could correctly identify the writer's gender almost three-quarters of the time.[22] In a similar vein, a team of computer scientists developed a language-based algorithm for identifying a writer's gender, claiming approximately 80 percent accuracy.[23]

A third way of assessing gender differences in written language is the use of standardized achievement tests. In the United States, the best-known yardstick of children's academic achievement is *The Nation's Report Card*.[24] Over the years, girls have consistently outpaced boys on the writing component of the test. The 2002 study reports that for students tested in grades

four, eight, and twelve, females outscored males, with the gap between genders being greatest in twelfth grade.

• • •

Studies of traditional spoken and written language clearly suggest that gender
influences both the reasons people use language (the "social" or "involved,"
versus "informational" dimension) and the standards to which their language adheres. The question now is whether gender distinctions surface in
computer-mediated communication.

Gender and CMC

The early days of online communication were marked by optimism that the
new technology would be a social leveler. If on the Internet nobody knows
you're a dog, no one knows your real age, social status, or gender either. Studies done in the late 1980s and early 1990s by Lee Sproull and Sara Kiesler
suggested that computer conferencing enabled people on the lower rungs of
organizational hierarchies to contribute more actively to decision-making processes in online meetings than happened face-to-face. Moreover, some female
students at then male-dominant Carnegie Mellon University reported feeling
more comfortable asking questions of their male faculty via email than in face-
to-face meetings during office hours.[25]

Over time, CMC researchers began to realize that online communication
hardly guaranteed either social or gender equity. Rather, as Susan Herring
has demonstrated, online dynamics often replicate offline gender patterns:

> The linguistic features that signal gender in computer-mediated interac-
> tion are much the same as those that have been previously described
> for face-to-face interaction, and include verbosity, assertiveness, use
> of profanity (and rudeness), typed representations of smiling and laughter,
> and degree of interactive engagement.[26]

Herring has also examined the discourse dynamics of online college-student
conversations, looking at both synchronous and asynchronous one-to-many
platforms. In both instances, Herring reports gender asymmetries. On asynchronous discussion lists and newsgroups, males typically dominate conversion:

> males are more likely to post longer messages, begin and close discus-
> sions in mixed-sex groups, assert opinions strongly as 'facts,' use
> crude language (including insults and profanity), and in general mani-
> fest an adversarial orientation toward their interlocutors

whereas females

> tend to post relatively short messages, and are more likely to qualify
> and justify their assertions, apologize, express support of others, and in
> general, manifest an 'aligned' orientation toward their interlocutors.[27]

In one-to-many synchronous CMC forums, gender roles were somewhat more balanced. In chat, for instance, there was more equal participation, as measured by number of messages and message length. However, gender differences (and often inequalities) still pervaded chat and social MUDS or MOOs. Males used more aggressive and insulting speech, whereas females typed three times as many representations of smiles or laughter. Male discourse was oppositional and adversarial, while female conversational style was aligned and supportive.[28]

What are the take-away lessons from Herring's studies? One is that at least in mixed company, women don't come across in one-to-many CMC as particularly loquacious. Another lesson is that the language women did use displayed the linguistic characteristics of "social" or "involved" communication (as opposed to "informative") that we earlier found both in face-to-face speech and in writing.

As a CMC medium, IM differs from Herring's scenarios in several important ways. IM is one-to-one communication. In IM, the conversational partners nearly always know each other, often quite well. (In one-to-many forums, you can generally participate anonymously or with a camouflaged identity, and many of the other participants may be strangers.) And finally, in collecting data from IM conversations, it's easy to gather samples from same-sex conversational pairs, facilitating the study of gender issues.

• • • THE IM STUDY

In spring 2003, my students and I explored how undergraduates (or very recent graduates) at American University were using IM with their friends. The version of IM we selected was America Online's free downloadable program known as AIM (AOL Instant Messenger), since nearly all students on campus seemed to be using it at the time. A group of student experimenters initiated IM conversations with peers on their AIM buddy lists. Everyone was given the opportunity to edit out any words or turns they wished to delete (an option rarely taken), and user screen names were anonymized. Student experimenters then electronically forwarded the IM conversation files to a project web site.

A Bit More Terminology

In writing this book, I vowed to keep technical jargon to a minimum. While the current chapter risks violating this goal, my hope is that by having a few terminological pegs upon which to hang our analysis of IM, we can better understand how IM conversations actually work.

Here are the terms we need to know:[29]

Transmission Unit: an instant message that has been sent
    e.g., Max: hey man

Utterance: a sentence or sentence fragment in IM
    e.g., Susan: Somebody shoot me! [sentence]
    e.g., Zach: if the walls could talk [sentence fragment]

Sequence: one or more IM transmissions sent seriatim by the same person
    e.g., Max: hey man
    Max: whassup
    [this sequence equals two IM transmission units]

Closing: a series of transmissions (between IM partners) at the end of an IM conversation, beginning with one party initiating closure and ending with termination of the IM connection
    e.g., Sam: Hey, I gotta go [first indication that Sam will terminate the conversation]
        ... [subsequent conversational transmissions]
    Sam: I'm outta here [final transmission in conversation]

Utterance Chunking: breaking a single IM utterance ("sentence") into two or more transmissions
    e.g., Joan: that must feel nice
    Joan: to be in love
    Joan: in the spring
Note: Each of the transmission units making up the utterance is an utterance chunk.

Utterance Break Pair: two sequential transmissions that are grammatically part of the same utterance
    e.g., Allyson: what are you bringing to the dorm party
    Allyson: on Saturday?

The most fundamental notion here is the IM **transmission unit**. Think of a transmission unit as a clump of writing that one of the people in the IM conversation composes and sends. Sometimes that transmission corresponds to a full sentence, as in Susan's "Somebody shoot me!" Other times, the transmission may be just a piece of a sentence, as with Max's "hey man" or Zach's "if the walls could talk." A third possibility is that a transmission contains more than one sentence. Jill, for example, wrote "and the prof left—he forgot something in his office."

An **utterance** is essentially a formal name for a sentence—or a piece of a sentence. Some utterances are fully contained within a single transmission unit (as with "Somebody shoot me!"). Other times, the utterance is broken up ("chunked") into multiple turns. Max's **sequence** of two transmissions that make up a single utterance (think "sentence") is a good example:

    transmission unit 1:   hey man
    transmission unit 2:   whassup

The meaning of **closing** is obvious: Think of a closing as a long goodbye.

Utterance chunking is simply the process of breaking an IM utterance (aka sentence) into multiple transmissions. Each one of the transmissions can be thought of as an utterance chunk. But where in the utterance does the chunking occur? If the total utterance is "hey man, whassup", does the break into two transmissions always take place between "man" and "whassup"? Why not between "hey" and "man"?

This "where" question turns out to be more important than you might think. When we talked about Chafe's notion of intonation units in speech, we saw that intonation units can be recognized either prosodically (by rising or falling pitch, or by beginning with a brief pause) or grammatically (beginning with a conjunction such as *and* or constituting a single clause, containing a subject and a predicate, such as "Somebody shoot me!"). A fundamental question is whether IM utterances are broken into sequential transmissions at the same grammatical points as spoken utterances.

The final term gives us a convenient way to talk about the relationship between two chunks within an utterance. **Utterance break pair** refers to two sequential transmissions that are part of the same utterance, as in

    transmission unit 1:   what are you bringing to the dorm party
    transmission unit 2:   on Saturday?

When we get to analyzing how IM utterances are chunked into multiple transmissions, we'll be asking what the grammatical relationship is between, for example, "what are you bringing to the dorm party" and "on Saturday."

Questions about IM

With our terminological ducks now in a row, let's turn to the specific linguistic questions we posed about IM. These questions cluster into three broad categories: conversational scaffolding, lexical issues, and utterance breaks.

Conversational scaffolding deals with how a conversation is put together. For starters, we sized up the individual IM transmissions: How long were they? How many consisted of just one word? How many transmissions were there per minute? Next, we considered how transmissions were combined to form sequences: What was the longest sequence in each conversation? How many transmissions were there per sequence? And how common were sequences in the corpus? Finally, we looked at conversation length: How many transmissions did we find per conversation? How long did conversations take? And how long did it take to say goodbye?

The second broad category of analysis was the lexicon, that is the words and short phrases that serve as building blocks for IM conversations. Here, we focused on various types of shortenings: abbreviations, acronyms, and contractions. We also looked at emoticons. While we were at it, we tracked the level of accuracy in the way words were written: How often were words misspelled, and how frequently did people make a mistake and then correct the error in an immediately following transmission?

The third set of questions involved utterance breaks. We've already explained a bit about how utterance breaks work. Remember that our interest is in seeing where in a sentence the breaks occur and then comparing these IM break points with the breaks in face-to-face spoken language.

• • • GENERAL FINDINGS

We collected 23 IM conversations, containing 2,185 transmissions, made up of 11,718 words. There were 9 conversations between females, 9 between males, and 5 involving male-female pairs. Given our relatively small sample size, this was, in essence, a pilot study.[30]

Conversational Scaffolding

Taking a bird's-eye view, we first looked at conversational scaffolding: a profile of the IM transmissions, sequences and utterance chunking, and conversation length and closings.

Transmissions

The average transmission (that is, a single IM that was typed and sent) was 5.4 words long. Averages, however, can be deceiving. Some of the transmissions were quite lengthy—the longest being 44 words. Others were really short. In fact, one out of every five transmissions was only a single word. Still, averages are hard data, and we need them for comparisons with other people's research.

Is 5.4 words long or short? In their contrastive analysis of spoken and written language, Wallace Chafe and Jane Danielewicz found that informal spoken conversational intonation units averaged 6.2 words, while academic lectures came in at 7.3 words. Moving to writing, Chafe and Danielewicz divided up prose according to "punctuation units" (essentially clumps of language set off by punctuation marks). The punctuation unit for traditional letters averaged 8.4 words, and written academic papers averaged 9.3 words.[31] At 5.4 words, our IM transmissions more closely resembled informal face-to-face speech than letters or academic works.

Another way of thinking about length is in terms of time. Because all the IM transmissions were time-stamped (a feature available through AIM), we could calculate not just how long each conversation lasted but how many transmissions were sent per minute. The average—barely 4 transmissions a minute—seemed low, considering how few seconds it takes to type 5 or 6 words and send the message. We knew that students could type more than 21.6 words per minute (5.4 times 4). Besides the time needed to read incoming messages, what else was going on?

To help find out, we analyzed the time gaps between IM transmissions. While many transmissions followed closely on the heels of the preceding message (this is, after all, supposed to be an "instant" medium), we also found a sizable number of long pauses between transmissions. Applying what we learned in the last chapter about multitasking, it's highly likely that students were busy writing papers, looking for cheap tickets to Europe, chatting face-to-face, or managing other IM conversations at the same time they were tapping out these 21.6 words.

Sequences and Utterance Chunking

Another reason that IM transmissions were, on average, relatively short is because so many IMs are written seriatim, together making up the equivalent of a sentence. Sequences turn out to be quite common. Nearly half the sample consisted of sequences of two or more transmissions. While some sequences contained only two or three transmissions seriatim, the longest sequence was

18 successive transmissions. (Talk about not being able to get a word in edgewise!)

We then looked more closely at the transmission sequences to see which ones contained distinct utterances, such as

transmission 1: i'm sorry [utterance 1]
transmission 2: if it makes you feel any better, i'm being held captive by two of Julie's papers [utterance 2]

and which constituted pieces of larger sentences, as in

transmission 1: in the past
transmission 2: people have found stuff under the cushions [together, a single utterance]

Eliminating the one-word transmissions (which were almost never part of multitransmission sequences), roughly one-sixth of the remaining transmissions in the data were part of an utterance break pair (that is, two sequential transmissions that are components of the same larger sentence). In a moment, we'll look at these break pairs in more detail.

Conversation Length and Closings

On average, the IM conversations were fairly lengthy: more than 93 transmissions apiece and almost 24 minutes long. In reality, IM conversations show enormous variety, ranging from quick three- or four-transmission volleys to sessions stretching over more than 200 transmissions and exceeding an hour.

We also examined how people ended their conversations. From the first indication that one of the partners intended to sign off up until actual closure, people took an average of 7 transmissions and roughly 40 seconds. Much as in face-to-face spoken encounters,[32] terminating an IM conversation can be a drawn-out process. Here's an example:

Gale: **hey, I gotta run**
Sally: Okay.
Sally: I'll ttyl?
Gale: **gotta do errands.**
Gale: **yep!**
Sally: Okay.
Sally: :)

Gale: **talk to you soon**
Sally: Alrighty.

Lexical Issues

Abbreviations, acronyms, contractions, and emoticons, along with spelling mistakes and self-corrections, are all lexical issues. Our findings hardly mirror the image of IM presented in the popular press.

Abbreviations

In tallying IM abbreviations, we included only what we might call electronically-mediated communication abbreviations—abbreviations that appear to be distinctive to online or mobile language. Excluded were forms that commonly appear in offline writing (such as *hrs* = *hours*) or represent spoken usage (*cuz* = *because*). Admittedly, the line is sometimes difficult to draw. For instance, *b/c* for *because* was included in the tally of EMC abbreviations, whereas *prob* for *problem* or *convo* for *conversation* was not.

Abbreviations proved to be quite sparse. Out of 11,718 words, only 31 were EMC abbreviations:

bc (also *b/c*) = *because*	5
bf = *boyfriend*	2
cya = *see you*	7
k = OK	16
y? = *why*	1

Acronyms

Again, we tabulated only acronyms that appear to be distinctive to EMC. That meant excluding acronyms such as *US* = *United States* or *TA* = *teaching assistant*, which are part of common offline speech and writing.

In all the conversations, there were only 90 EMC acronyms:

brb = *be right back*	3
btw = *by the way*	2
g/g (also *g2g*) = *got to go*	2
LMAO = *laughing my __ off*	1
lol (also *LOL*) = *laughing out loud*	76

OMG = oh my god	1
ttyl = talk to you later	5

LOL was the runaway favorite, but the term didn't always indicate the humorous response suggested by the words "laughing out loud." Rather, *LOL*, along with *heehee* or *haha* (both also common in IM), were sometimes used as phatic fillers, the equivalent of *OK*, *cool*, or *yeah*:

Mark:	i've got this thing that logs all convos [=conversations] now
Jim:	really?
Jim:	why's that
Mark:	i have ever [=every] conversation i've had with anybody since the 16th
Mark:	i got a mod [=module] for aim [=AIM], and it just does it
Mark:	i'm not sure why
Jim:	lol
Jim:	cool

Contractions

If IM is like speech, we would expect contractions to pop up wherever the language permits (such as *I'm* instead of *I am* or *he's* rather than *he is*). But that is not what we found.

Out of 763 cases in which the participants could have chosen a contraction, they did so only 65 percent of the time. Compare this situation with casual speech. For a class project, some of my students once tallied how often contractions were used in a sample of college students' informal conversation. The answer: roughly 95 percent of the time. The surprising thing about uncontracted forms in IM is that they occur as often as they do, bringing to the messages a more formal tone than we usually associate with IM.

Emoticons

Emoticons were also in short supply—a total of only 49:

:-)	= smiley	31
:-(	= frowny	5
O:-)	= angel	4
:-P	= sticking out tongue, with nose	3
;-)	= winking	2

:-\ = undecided	1
:-[= embarrassed	1
:P = sticking out tongue, without nose	1
:- = [probably a typographical error]	1

The odds-on favorite among emoticons was the smiley: 31 out of the 49. Yet not everyone used emoticons. Just 3 subjects accounted for 33 of the emoticons.

Spelling and Self-Correction

Only 171 words were misspelled or lacked necessary punctuation. That averages out to only one error every 12.8 transmissions. Not bad. (My students' essays sometimes show poorer spelling.) Here are the types of errors and (rounded) percentages of the 171 words in which they appeared:

missing apostrophe (e.g., *thats*):	37%
letter added, omitted, error (e.g., *assue* for *assume*; *coliege* for *college*)	33%
letter metathesis (e.g., *somethign*):	21%
other (e.g., homonyms—*your* for *you're* phonetic spelling—*dido* for *ditto* compounding mistakes—*over drew* for *overdrew*)	8%

More than one-third of the errors came from omitting an apostrophe in a contraction (such as *thats*) or a possessive form (*Sams*). Another third were garden-variety spelling mistakes—adding or omitting letters, or using the wrong letter. Some mistakes probably came from sloppy typing, as did the 21 percent of errors involving metathesis (switching letter order, as in *somethign* for *something*).

In about 9 percent of the cases, the person noticed the problem and fixed it (or tried to) in the next transmission. For instance, when a subject had typed *awway* (and sent off the IM), he corrected it to *away* in the following transmission. Self-corrections didn't follow a clear pattern. Changing *awway* to *away* probably wasn't necessary for the recipient to interpret the original message. Other errors that were not corrected did challenge intelligibility. (I for one would have benefited by seeing *feidls* corrected in the next transmission to *fields*.)

Out of the 15 self-corrections, none involved adding in a missing apostrophe. As I'll suggest later on, the English apostrophe might be an endangered species. But we should also add that the only self-corrections we can report on were those visible in the conversational record. Several researchers have observed that users sometimes edit their IM messages before sending them.[33] Yet unless you archive every keystroke the IM participants make or you train a video camera on the computer screen of the person typing (which Gloria Jacobs actually did),[34] it's impossible to record accurately all the corrections that go into IM conversations.

Utterance Breaks

Finally, utterance breaks. The analysis here is a bit more complex but worth following. To give the game away: Males and females break their utterances differently, and these differences offer important clues about whether IM should be thought of as speech or writing.

For the utterance break analysis, we looked at 18 conversations: 9 between females and 9 between males. The female pairs sent more individual transmissions (1097 versus 767). However, males were significantly more likely to break their utterances into chunks, each of which was sent as a separate transmission.[35] Here's the summary (with the total percentages rounded):

	Total Break Pairs	Total Transmissions in Corpus	% of Total Transmissions
Females	84	1097	13%
Males	105	767	23%
Total	189	1864	17%

Grammatical Make-Up of Break Pairs

The next step was to code each of the 189 break pairs by figuring out the grammatical relationship between the first transmission and the second. Take the break pair sequence

transmission 1: what are you bringing to the dorm party
transmission 2: on Saturday?

The second transmission ("on Saturday") functions as an adverbial prepositional phrase, modifying the sentence in the first transmission ("what are you bringing to the dorm party"). The coding scheme I devised looked at the grammatical structure of the *second* member of the break pair, in relationship to the first member. In our example, that means I looked at the grammatical function of "on Saturday."[36]

Grammatical Analysis of Utterance Break Points

Here are the results of our grammatical coding (again, the percentages are rounded):

Grammatical Type	Females	Males	Total
Conjunctions and Sentences or Phrases Introduced by Conjunctions	48%	69%	59%
Independent Clauses	23%	9%	15%
Adjectives and Adjectival Phrases	7%	8%	8%
Adverbs and Adverbial Phrases	12%	6%	9%
Noun Phrases	9%	6%	7%
Verb Phrases	1%	2%	2%

Conjunctions are the primary device for breaking utterances into multiple transmissions. Out of 189 break pair sets, 112 (nearly 60 percent) began the second transmission with a conjunction. Of these 112 cases, 89 used a coordinating or subordinating conjunction to introduce a sentence (such as "*and* she never talks about him" or "*if* I paid my own airfare/"). The remaining 23 were conjunctions introducing a noun phrase ("*or* circleville") or verb phrase ("*and* had to pay back the bank"). If we separate out all the coordinating conjunctions (such as *and* or *but*) from all the subordinating conjunctions (for instance, *because* and *although*), we discover that more than four out of five of IM transmissions that appeared as the second member of an utterance break pair (and that began with a conjunction) used a coordinating conjunction. That's a lot of coordinating conjunctions.

The next most prevalent grammatical type for beginning the second transmission in a break pair was independent (sometimes called "main") clauses (for example, "that's all I'm saying"). Of the 189 break pairs, about 15 percent constituted independent clauses. Grammatically, independent clauses are also sentences (or sentence fragments). If we add together conjunctions introducing sentences with the independent clause category, we account for roughly 62 percent of the total 189 utterance breaks. Clearly, sentence units (whether or not preceded by a conjunction) constitute an important pattern for constructing IM sequences.

The remaining cases of second transmissions in a break pair were largely adjectives ("completely harmless"), adverbs ("on Saturday"), or nouns ("radio station"). Then there were a couple of stragglers, plus one lone example in which the second element was a full-fledged verb phrase:

transmission unit 1:   and then Pat McGee Band
transmission unit 2:   perform like 7

This outlier bears more attention.

Intuitively, an utterance break between the two main constituents of a sentence—the noun phrase subject and verb phrase predicate—seems a natural place to anticipate finding chunking in IM conversations. What's odd is that this pattern occurred only once in all the IM conversations. This fact will prove relevant in deciding if IM more closely resembles speech or writing.

• • • THE GENDER QUESTION

Now that we've seen the overall make-up of our IM sample, what happens when we reexamine the conversations by gender? We'll zero in on those features where gender seems to make a difference.

Conversational Scaffolding

Transmissions

The prize for the longest IM transmission, in words, goes to the females: 44 words. (The longest male transmission was 34 words.) If we average the longest transmissions in each of the 9 female conversations and then in each of the 9 male ones, females again have the upper hand. The female "longest transmissions" averaged almost 28 words, while the males averaged not quite 20 words. In short, females writing to females account for more of the lengthy transmissions than males writing to males.

Sequences and Utterance Chunking

Females used considerably more multitransmission sequences than did males. However, when we hone in on just those sequences that chunk sentences into multiple transmissions, the balance shifts. Males were almost twice as likely to carve up sentences into sequential transmissions as females were.

Conversation Length and Closings

Female–female conversations were roughly a third longer (in both number of transmissions and time on the clock) than male–male conversations. Females averaged almost 122 transmissions per conversation, lasting an average of 31 minutes. Males averaged 85 transmissions per conversation, with conversations averaging only 19 minutes in duration.

Lexical Issues

Gender was irrelevant for all lexical categories except contractions and emoticons, but here the differences were palpable.

Contractions

We reported that students used contractions 65 percent of the time they had the option. But usage differed significantly across gender lines. While males used contracted forms more than three-quarters of the time (77 percent), females used contractions in only 57 percent of the possible cases.

Emoticons

If males were more likely to use contractions, females were the prime users of emoticons. Three-quarters of the 16 females in the study used one or more emoticons. Of the 6 males, only 1 used emoticons.

Utterance Breaks

Finally, back to utterance breaks. We already know that males used a higher proportion of utterance breaks than females did. But where did those utterance breaks appear? Males were significantly more likely than females (69 percent versus 48 percent) to begin the second transmission in a break pair with a conjunction. And females were significantly more likely than males (23 percent versus 10 percent) to chain together related sentences. To interpret what these differences mean, we turn to the bigger question of whether IM is more a spoken or written medium.

• • • IM AS SPEECH OR WRITING

Let's start with a score sheet to see how IM stacks up in comparison with face-to-face speech or conventional writing:

	Similar to Face-to-Face Speech	Similar to Conventional Writing
GENERAL DISCOURSE SCAFFOLDING		
• average turn length	yes	no
• one-word utterances	yes	no
• conversational closings	yes	no
LEXICAL ISSUES		
• use of contractions	somewhat[37]	somewhat
• EMC abbreviations, acronyms	no[38]	somewhat
• emoticons	yes[39]	no
UTTERANCE BREAKS (UB)		
• frequency of chunking utterances into multiple sequential transmissions	yes	no
• 2nd member of UB pair begins with conjunction ·	yes	no
• 2nd member of UB pair begins with coordinating conjunction	yes	no
• 2nd member of UB pair begins with independent clause	somewhat[40]	yes

Outside of a few lexical issues, plus utterance breaks involving independent clauses, the score sheet clearly suggests that IM more closely resembles speech than writing. However, since a lot of the analysis hinges on utterance breaks, we need to take a closer look at the relationship between IM utterance breaks and the way spoken discourse works.

On a number of points, the IM data and Chafe's findings for speech seem strongly congruent. Average lengths for both the IM transmission units and Chafe's intonation units are relatively short (IM: 5.4 words; intonation units for informal speech: roughly 6 words).[41] In both cases, coordinating conjunctions commonly initiate a new transmission or intonation unit. What's more, in both instances, new units are sometimes made up of independent clauses.[42]

On other measures, the comparison fails. In the IM data, grammatical breaks between adjectives and nouns were infrequent (a total of 5 examples out of 189), and breaks between noun phrases and verb phrases were downright rare (only one case). Chafe, however, reports multiple instances in spoken language in which a pause occurred between an adjective and a noun, or falling intonation separated a noun phrase and a verb phrase. For example,

adjective ("the")—noun ("road"):
  and spilled the pears all over the [pause] road [43]
noun phrase ("the picker")—verb phrase ("was picking the pears"):
  where the picker [falling intonation] was picking the pears[44]

The Significance of Gender in IM

Earlier in the chapter, we talked about "involved" (or "social") versus "informational" discourse, and about standard versus non-standard usage. We saw that female language (both spoken and written) is more likely to have an involved or social function, while male communication tends toward being informational. Similarly, female language is, on the whole, more standard than that of males.

How do these findings about traditional spoken and written language match up with the IM data?

In IM, females were more "talkative" than males. Women had the longest individual transmissions, had longer overall conversations, and took longer to say goodbye. In their study of informal essay-writing, Mulac and Lundell found that females used longer sentences than did males. It's possible, then, that our IM findings reflect a female *writing* style rather than a female *speech* style.

Another measure of the social function of EMC conversation is use of emoticons. In our data, females were far more likely to use emoticons than males. Herring reported a similar finding in her study of one-to-many synchronous communication. Since emoticons can be interpreted as visual cues used in lieu of prosody or kinesics (which appear in face-to-face communication), their use in IM would suggest a more spoken than written cast.

The finding that females used fewer contracted forms than males suggests that women have a greater tendency to treat IM as a written rather than spoken medium. Anecdotal evidence from my students suggests they have all been taught that contractions should not be used in formal writing, although they don't always follow this rule.

Rich Ling's studies of Norwegian teenagers and young adults show that females used more standard punctuation and capitalization in their text messaging on mobile phones than males.[45] While we didn't completely code the IM data for punctuation and capitalization, informal analysis suggested that females more strictly adhered to the rules than males—again, an indication that female electronically-mediated communication more closely follows written norms than does male language.

The utterance-break data shed yet more light on the question of how gender shapes IM as spoken or written discourse. Males were significantly more likely to break up sentences into multiple IM transmissions than were females. Males were also significantly more apt than females to begin the second member of an utterance break pair with a conjunction, while females were significantly more prone than males to begin the second member of such a break pair with an independent clause. The conjunction pattern is more commonly found in speech, whereas the independent clause pattern is more characteristic of writing. While the intonation units that Chafe and Danielewicz analyzed in face-to-face speech began with a coordinating conjunction 34 percent of the time, only 4 percent of the punctuation units in their academic writing sample began this way.[46]

Synthesizing all these gender-based distinctions, we need to rethink our earlier tentative conclusion that IM looks more like speech than writing. Breaking out our initial score sheet by gender suggests a more nuanced story:

	Similar to Face-to-Face Speech		Similar to Conventional Writing	
	Males	Females	Males	Females
GENERAL DISCOURSE SCAFFOLDING				
• conversational closings		longer	shorter	
LEXICAL ISSUES				
• use of contractions	more frequent			less frequent
• emoticons		more frequent[47]	less frequent	

	Similar to Face-to-Face Speech		Similar to Conventional Writing	
	Males	Females	Males	Females
UTTERANCE BREAKS (UB)				
• frequency of chunking utterances into multiple sequential transmissions	more frequent			less frequent
• 2nd member of UB pair begins with conjunction	more frequent			less frequent
• 2nd member of UB pair begins with independent clause	less frequent			more frequent

While male IM conversations have a great deal in common with descriptions of face-to-face speech, female IM conversations more closely approximate conventional writing patterns. The only two exceptions to this generalization are conversational closings and use of emoticons, both of which were more pronounced among females than among males, and both of which are more analogous with traditional spoken than written communication.

• • • BOTTOM LINE: ARE INSTANT MESSAGES SPEECH?

The simple answer: no, even though there are enough speechlike elements (especially in male IM conversations) to explain why it seems so natural to talk about IM "conversations" and not IM "letters." Just as we commonly speak face-to-face while engaging in additional activities (walking down the street, doing the dishes), young people are typically doing something else (online, offline, or both) while conducting IM conversations. As with speaking (and unlike writing), IM conversations are not generally targets of someone's normative eye or red pen. The goal of an IM conversation is to get your message across (boredom, empathy, arranging to meet tomorrow, gossip), not to produce an entry for an essay contest.

At the same time, because IM is overwhelmingly informal, by the time users reach college, they tend not to put a lot of effort into monitoring what they write. Younger teenagers may care about looking cool by only using *U* for 'you.' (In fact, the fifteen-year-old son of a colleague admitted that he intentionally included abbreviations so he wouldn't look like a nerd.) On the other hand, my college students seem to have neither time for nor interest in such linguistic posturing. For them, IM is far more pragmatic.

So why does an informal medium like IM assume some dimensions of more formal, written language? The answer reflects what the philosopher John Dewey called habit strength. By the time they reach college in America, today's freshmen have probably been using ten fingers on a computer keyboard for more than a decade. Much of their typing is for school work, which is apt to be subjected to academic scrutiny. Acronyms, abbreviations, and contractions are no more welcome than poor spelling. After a while, school children get the hang of what is required. With years of repeated practice, their fingers tend to go on automatic pilot. It's not surprising to see school-appropriate writing habits crop up in IM, which is produced on the same keyboard as those formal school assignments—and sometimes, at the same time. Since girls, on average, produce better writing than boys do in the K–12 years, it makes sense to expect that female IM conversations will reveal heightened standards, including fewer contractions, fewer sentences chunked into multiple transmissions, and fewer sentence breaks involving a conjunction, in comparison with males.

Speech or writing? Some of both, but not as much speech as we've tended to assume. What's more, gender matters.

5

• • • My Best Day

Managing "Buddies" and "Friends"

Walk into an average college dormitory, and wander down the halls. More often than not, the stark cinderblock or plasterboard is relieved by decorations—travel posters, likenesses of Leonardo DiCaprio, or maybe flyers advertising a used computer or bike for sale. At each room, you commonly find a corkboard or whiteboard, on which messages of all ilk can be written either by or for the occupants. Sometimes personal photographs are tacked up, or maybe lyrics from a favorite song or a quotation from Marx or Monty Python.

These bulletin boards let students post timely messages for friends: "I've already gone to dinner" or "Wanna see the flick tonight?" They also serve a much richer function: enabling people to craft a presentation of self to a limited circle of friends (plus those with access to the hallway).

Fifty years ago, the sociologist Erving Goffman introduced the notion of "presentation of self" as a formal social construct. Goffman argued that people consciously or unconsciously present themselves to others as if they were actors on a stage: Do I want to appear assertive? Vulnerable? Sophisticated? Available?[1] Goffman's work faded from prominence in the final decades of the twentieth century, but his notion of presentation of self has found a new audience among Internet researchers who study how we use online media in establishing social rapport.

If personal bulletin boards are tangible devices for communicating with classmates who are physically proximate, virtual platforms such as instant messaging and social networking sites (for instance, Facebook or MySpace) offer additional outlets for conveying information or socializing. But online sites are also places for constructing images of how you wish others to perceive you. In choosing the title for this chapter, I gratefully borrowed a phrase one student used in discussing how she thought about her Facebook page. "My best day," she said, meaning she could stage herself as she wished to be seen by her friends.

It is this notion of staging, of "my best day," that we'll be exploring in this chapter. While the last chapter looked at IM as a one-to-one medium, this chapter widens the social lens to one-to-many communication. The two venues we consider are away messages in instant messaging and then Facebook—both platforms for orchestrating social relationships.

• • • SLEEPING…OR AM I ☺: AWAY MESSAGES IN IM

"Out of sight, out of mind." While years of teaching have convinced me that today's students know fewer and fewer aphorisms, the import of this one is readily understood by teenagers and young adults, as evidenced by the way they manipulate electronically-mediated communication. Yes, computers and mobile phones are used when people are out of one another's sight, but the art comes in not being "out of mind." Mobile phones are ideal for fifteen-second calls ("It's me. How you doing? I'm fine. Gotta go. Bye.") or comparably short text messages ("Hey. Lov ya"). With computers, away messages often play a similar role. Ostensibly, people posting an away message are not at their computer. But their presence lingers through their words.

How Away Messages Work

Historical Note: The IM platform we're about to describe is America Online's Instant Messenger, commonly known as AIM—but vintage late 2002. This was the system running when we collected the away-message data discussed in this chapter. Over time, features on IM platforms have mushroomed to include voice and video, multiparty conversations, and the ability to send IMs to users not currently logged on.

Away messages are part of a broad suite of IM functions enabling users to send synchronous messages to individuals but also to "present" themselves to members of their buddy list or anyone knowing their screen name. These forms of presentation include screen names, profiles, buddy icons, fonts and colors, and away messages.

Users can manipulate all these functions. Selection of screen names can be a creative act (such as the choice of "Swissmiss"—also the name of a hot-chocolate mix—by an American who had lived in Switzerland). Profiles and buddy icons enable users to craft a persona (real or imagined) through choice of web site links, quotations, and avatar imagery. Similarly, users have the option of customizing fonts and colors when constructing text for IMs, profiles, or away messages.

Away messages were originally designed to enable AIM users who were still logged on to their computers but not physically sitting at their machines to alert friends not to expect immediate replies to IMs. During your absence, an away message creates a social link with other members of your messaging circle. As one female undergraduate put it, "Even if they are not chatting [on IM], you can still know all about someone's life by reading their away messages."

Think of away messages as a form of "onstage" behavior in contrast to IM conversations, which are more "backstage" activity (Erving Goffman's terms, again). Gloria Jacobs argues that among American teenage girls, "the backstage conversations [that is, IM] are where alliances are formed, problems are discussed and solved, and plans are made beyond the hearing of others... [while] the onstage places [away messages] are where alliances are declared and social positions and presence are established."[2]

AIM users know that a member of their online social circle has posted an away message by looking at the buddy list appearing on their screen. The list indicates which members of the group are currently online plus which ones (of those logged on) have posted away messages. A (virtual) yellow piece of paper next to a buddy's screen name indicates that the person has posted such a message. Click on the piece of paper, and you can view what is written. AIM provides a default away message ("I'm away from my computer right now"), and hundreds of web sites list thousands of sample texts.

Knowing how away messages function in principle is one thing. But my students kept telling me that the actual use of away messages was not what the name would lead you to expect. We decided to explore.

• • • THE IM AWAY-MESSAGES STUDY

In fall 2002, we gathered a collection of IM away messages that had been posted by friends on their AIM buddy lists. Over a two-week period, we compiled five different messages from each of 38 people (half male, half female), giving us 190 messages. Several students also interviewed their subjects, eliciting the writers' rationales for constructing away messages.[3]

The messages yielded an array of styles and moods, a good deal of humor, and a substantial display of personal information. While people sometimes recycled their own away messages (since they can be saved), no one resorted to AIM's default away message or to public messaging sites. In our class discussions of the project, students also volunteered messages they had used themselves or encountered in their experience with IM.

Message Length and Gender

Message length varied enormously across individual subjects, ranging from 1 word to more than 50. Among females, the average length of an away message was 12.3 words. For males, it was 13.3.

Compared with IM transmissions, away messages are fairly long. Why? Part of the explanation may be that away messages are one-shot deals, while several IM transmissions are often sent seriatim (with their cumulative word total longer than the transmission average of 5 or 6 words). However, the more interesting part of the answer lies in the role away messages play in students' social lives.

Content Analysis

The messages naturally clustered into two main categories. The first served to convey information or start a conversation, while the second essentially provided entertainment. Each category then had subdivisions, which we'll get to in a moment.

While coding the data, we immediately noticed there was often a gap between the surface, or overt, meaning of an away message and the mood, tone, or ulterior motive involved. Think of the difference between denotative and connotative meaning. Denotatively, a message might indicate that although I still am logged on to IM, I've gone to the library, am not at my computer, and therefore won't be responding to messages you might send. But if the message says I am *In the bowels of hell. . . . or what some would call the library,* I have connotatively relayed far more information.

It turned out to be easiest to analyze the messages according to the overt text and then explain the actual communication functions. A number of messages overlap categories (especially where humor or a quotation is involved). These samples are included under the particular category being illustrated, though we should remember they may serve other functions as well.

Informational/Discursive Messages

We subdivided the first cluster of messages into four categories:

- "I'm away"
- initiate discussion or social encounter
- convey personal information (about yourself, your opinions, your sense of humor)
- convey detailed information to selected other(s)

"I'm away" messages overtly declare their authors are away from their computers and therefore not available to respond to IMs despite still being logged on to AIM. In some cases (such as a message that just said *"out"*), the overt message matched up with the communicative intent: "I really am away." In many other cases, however, the message carried added meaning. Here are some examples of messages that convey further information or intent:

a. ITINERARY: *"voter registration, peace corps meeting, class, class, choir, dinner, dorm council"*

  Itinerary messages spell out the sequence of activities in which the person posting the message will be engaged. While the level of detail may seem unnecessary, students noted the social usefulness of informing friends how they are spending their day. These specifics enable members of an online cadre to continue a conversation stream ("So how was your test?") rather than needing to begin a dialogue from scratch ("What did you do today?") when they resume IMing or encounter one another face-to-face.

b. RANDOMLY SELECTED MESSAGE: *"cleaning my room"* (when actually the person was not)

  Some AIM users care less about laying out their agendas than getting out the word that they will be unavailable. For this purpose, people sometimes grab whatever message from their saved arsenal they happen upon first. As one student explained, it's irrelevant if you are actually cleaning your room or off at a class. In either event, you're not responding to IMs.

c. REMAINING IN THE LOOP: *"Not here . . . Please leave me a message! Thanks."*

  These postings constitute requests for a message to be waiting when the individual posting the away message returns to active use of AIM.

This function is similar to telephone voicemail ("I can't take your call now. Please leave a message.").

d. LURKING/FILTERING: *"Maybe I'm doing work... maybe I'm not... the question of the night"*

Another socially motivated function of "I'm away" messages is monitoring incoming IM traffic, allowing people to decide which messages to respond to and which to ignore. An away message such as *"Sleeping... or am I ☺)"* or *"Maybe I'm doing work... may I'm not"* signals buddies there's some chance their IMs will be read (and responded to) immediately, but the recipient of such an IM is not obligated to do so. Apparently a number of college students post "I really am away" messages (such as *"out"*) when they are actually sitting at their computers. This ruse enables them to selectively ignore incoming IMs but to commence IMing if someone on the buddy list sends an interesting message.

e. INTENTIONAL MISREPRESENTATION: *"dinner with Mark and dancing all night"* (actually the person was alone in her dorm room, watching TV)

These socially motivated "I'm away" messages enable people to construct a self-image through creative license. Student researchers reported cases in which friends posted away messages detailing socially impressive activities (such as an elaborate date with a desirable partner) when they knew the person hadn't gone out for the evening. Since computer-mediated communication invites construction of new identities (for age, gender, personality, nationality, and the like), it's hardly surprising to find fabrication of activities in away messages.

Given the ostensible function of away messages (to say the writer can't be reached on IM), why not sign off from AIM if you won't be responding? Part of the answer lies in the technology. The default setting of AIM triggers a sound whenever people log on or off, alerting everyone on their buddy list. While the default sound can be turned off, a visual icon still appears, showing a door opening or closing. Students suggested that such an intrusion was socially "too loud." Not only are you noisily announcing your presence (when you log on) but you are inviting a deluge of IMs. Commenting on the "lurking/filtering" function (or use of an "I really am away" message when you're not), students again noted the importance of social politeness. If you are on AIM but want to converse only with specific people, posting an away message and then only responding to IMs selectively is a way of not hurting the feelings of those whose IMs you ignore.

A second group of informational/discursive messages invite communication now or in the future: by phone, text message, IM, or face-to-face. Some examples:

a. REACH ME THROUGH A DIFFERENT MEDIUM: *"since I am never around try TEXT MESSAGING me! Send it to [phone number]. I wanna feel the love!"*

These messages are direct attempts to remain in the social loop. The primary benefactor may be the person posting the away message ("I wanna feel the love!"). Other times, posters report a sense of obligation to their buddies. As one subject said, "I feel like I should be accessible. My cell phone is always attached. I don't want my friends to think there's a time when they can't reach me."

b. LET'S CHAT ONLINE: *"Please distract me, I'm not accomplishing anything"*

Other messages invite buddies to IM the person posting the away message when he or she is working at the computer. This "boredom" function looks like an oxymoron, but it turns out to be an efficient way to broadcast a request to many readers in the hope that someone replies. As with "lurking/filtering" messages, individuals posting away messages that solicit online conversation can select the incoming IMs to which they wish to respond.

Away messages are also used to convey personal information about yourself, your opinions, or your sense of humor. Here are several examples:

a. CURRENT ACTIVITY: *"Reading for once . . . the joy of being an English major is sooo overwhelming right now . . . (the sarcasm is very much intended)"*

b. OPINIONS: *"You have very little say in your fate or what will eventually befall you, but don't let that keep you from voting."*

c. SENSE OF HUMOR: *"I could easily be replaced with a dancing chimp . . . and at times I believe people would prefer the chimp (but then again, so would I)"*

More often than not, if there's an opportunity for humor, the user takes it.[4] For example, a student who left her IM program running while at dinner could have written "at dinner." Instead, she combined the "I'm away" function with personal wit, resulting in *"this chick needs filla"* (a play on the name Chick-Fil-A, an on-campus fast food shop).

The last group of informational/discursive messages broadcasts to your entire social group detailed information that's intended for a specific person (or persons). For example, one writer posted *"working at the multicultural bilingual center . . . Sam, we will hang later tonight!!! I promise!"* while another wrote *"Back in D.C. missing Houston very much . . . Suz, Jan and Rick, thank you for an amazing weekend!"*

Why do some AIM users choose the public forum of away messages to convey information seemingly targeted to particular members of their buddy list? One explanation lies in AIM technology. As of fall 2002, AIM users couldn't send IMs to people not signed on to the system. The only way to communicate (using AIM) was through away messages, which could be accessed when the intended reader returned online. Another explanation is rooted in attitudes regarding privacy. One student we interviewed made clear that he writes what he feels like writing and doesn't care who sees it. Other students said they generally communicated through IM with a tight circle of friends, all of whom would likely know the individuals named in an away message.

Sometimes the rationale for posting "private" messages is less friendship than public display. By addressing or referring to a significant other or a named friend in an away message, writers publicize their personal relationships, reminding members of their buddy list that they are the sort of person who has such friends. This public display function is similar to the use of "I'm away" messages containing intentional misrepresentation in order to impress others (such as claiming to be on a date when actually at home).

Entertainment Messages

Some away messages are primarily posted to entertain, using humor, quotations, or even song lyrics. Examples we collected include:

a. HUMOR: *"work rhymes with beserk and jerk. i lurk in the murk and do my work, ya big jerk. ok, I gotta go do some work now."*

b. QUOTATIONS: *"'Good breeding consists of concealing how much we think of ourselves and how little we think of other people'—Mark Twain"*

Why post away messages for the primary purpose of entertainment? Once again, the answer is grounded partly in the technology and partly in the social goals and expectations of American college students. Experienced users of email are familiar with signature files, which enable senders to automatically post at the end of their emails not only professional contact information but also pithy sayings or quotations. While IMs themselves have no signature files, "entertainment" away messages serve comparable functions, in essence providing a platform for self-expression.[5]

Many of my students—and the subjects whose messages they collected—perceived entertainment to be an essential component of the entire away-message enterprise. One person noted, "I like to make people happy with my messages." Another indicated that since she enjoyed friends' away messages that made her laugh, she tried to make her own words funny. A third said he likes to entertain people. Several others felt they had to justify themselves when their away messages were not funny or creative, typically explaining that they lacked time or energy to craft amusing postings.

How Senders View Away Messages

Away messages have multiple functions, not all of which are revealed by the overt form of the messages themselves. Interviews with some of our subjects provided helpful insight.

There was no consensus as to how much away messages should reveal or justify their author's whereabouts. Some users deemed it important to let potential conversational partners know why the person posting an away message was absent and how to locate that person. Others felt strongly that specifying their precise location was an invasion of privacy.

Similarly, away-message users differed regarding the appropriate length of a message. While some advocated (and sent) one word messages, others scoffed at the "laziness" of such writers:

> I don't appreciate/agree with people whose away messages consist of one word (such as "away," "sleeping," or "work"). I know these people are more interesting than that, and away messages can be indicative of your mood, your state of mind, and what you're doing at the time. The best ones can do all 3.

Some users consciously manipulated the away-message genre to serve individual needs. One interviewee noted that she only posts away messages when she is in her dorm room, working at her computer. (Her messages included the likes of *"Eating the souls of my fellow man"* and *"*sigh*"*.) For her, away messages were a way of expressing personal information (sometimes humorously) about her current situation, perhaps to generate conversation with people viewing her messages.

The flip side of using these messages to express your feelings is to craft messages that intentionally camouflage your state of mind. One subject commented that she posts quotations when she doesn't feel like talking or "giving away too much information." The same individual reported using self-deprecation (*"I could easily be replaced with a dancing chimp"*) when "it has been a long day" and she doesn't want to go into the details of why. Another woman revealed that she uses humor to mask her stress level in order not to bother friends with her troubles, but at the same time hinting that not all is well.

A comprehensive summary of how to construct an away message, posted in the AIM profile of someone on the buddy list of a member of the student research group, included this advice:

> it is important not to underestimate the value of a good away message. too much internet time is wasted by people reading mediocre/poor away messages. a few rules to go by:
>
> 1. no one word away messages—EVER
> 2. quotes/lyrics, unless appropriately timely, are a poor excuse for away messages and make the writer look like a hack
> 3. humor is the only way to go—i'm not looking for a deeper understanding of life, or a little tug on the heart strings from my instant messenger
> 4. don't leave your cell phone number. people aren't looking at your away message to contact you, they're looking at it cause they're bored out of their mind writing some paper

IM in the Age of Facebook

The subjects in our study had devised an intricate culture of away messages, capable of expressing a wide array of meanings. But like fashions in food or vacation spots, practices in electronically-mediated communication also change with the times. The away-message data were collected in fall 2002. Since that time, the online communication scene has evolved enormously. One of the biggest changes has been the proliferation of social networking sites, which enable groups of people to establish online social affinities using tools such as profiles and messaging that are in some ways reminiscent of IM. Among the American college crowd, the most important of these new platforms has been Facebook.[6]

• • • MY BEST DAY: FACEBOOK EVOLVES

By fall 2003, the first generation of social networking sites such as Friendster and Tribe.net were developing significant followings. Launched in beta form in fall 2002, Friendster had over 5 million registered accounts by early January 2004. On Friendster, users post profiles of themselves, write public testimonials about other users, and then browse the system in search of "Friends." As danah boyd explains, Friendster was created to compete with online dating sites such as Match.com. What made Friendster different was that rather than cruising blindly among the postings, you worked through friends of friends.[7]

American colleges and universities have long sought ways for their students to get to know one another. The venerable freshman "facebook" has been a fixture on college campuses for decades. Entries for each student typically include name, mug shot, date of birth, high school, hometown, and maybe college dorm address, potential major, and hobbies. Given students' growing technological savvy, various schools were now considering putting these facebooks online.

In November 2003, Mark Zuckerberg (a Harvard sophomore at the time) began doing programming for a social networking site specifically for Harvard students.[8] The design followed the lines of Friendster, but it also addressed some of the functions of traditional collegiate facebooks. On February 4, 2004, TheFaceBook.com went online. By the end of the month, more than three-quarters of the undergraduates at Harvard had signed up.

TheFaceBook (later renamed simply Facebook) really did spread like wild fire. It first launched in a handful of other schools (Stanford, Columbia, and Yale). By June 2004, there were forty institutions; in September, the number of users was up to a quarter-million. Within a year, Facebook had become the second-fastest-growing major site on the Internet—surpassed only by MySpace.

Facebook quickly swept across nearly all four-year colleges and universities in the United States. In September 2005, the site was opened to high schools—at first, without links to the college version but later merging the two. By the time the next academic year (2006) rolled around, the lid was taken off of Facebook, making it available to "corporate" and "regional" networks around the world—in essence, to anyone with an email address.

The Features on Facebook

Note: The profile that follows is based on Facebook as of fall 2006. New features continue to be added, such as "status" reports (on what a user is doing right now), along with virtual "gifts" and a growing number of applications from third-party developers. Because the list is a moving target, I focus here on the core features that have characterized Facebook for most of its history.

The main social-networking features of Facebook cluster into three categories: information about yourself, social affiliations, and online interaction.

Facebook also provides a collection of Privacy Settings allowing users to block the prying eyes of people they wish to keep out. Our description of Facebook only captures the highlights, not every last option, such as posting Spring Break plans or all the privacy permutations.

Information about Yourself

Like any traditional college facebook entry, Facebook opens with a picture. (If users choose not to post one, the site provides a question mark in its stead.) People also have the option of uploading photo albums, typically containing pictures of the page owner, along with his or her real-life acquaintances.

Most user information appears in the Profile. The basics include name, sex, birthday, and hometown, augmented by college-minded categories such as "Relationship Status" and "Political Views." Contact information options include AIM screen name, mobile phone number, mailing address, and web site. Then come the personal settings, where people can write about their activities and interests, indicate preferences in music (books, TV shows, movies), and offer up favorite quotations. Other settings include places for current and former academic information (field of study, high school attended), and employment details.

We know that online communication invites the use of assumed identities. In principle, Facebook is different, because at least until recently, you needed a .edu email address to register, which you possessed only if you were at a college or university. If you said you were Jaime L. Hernandez, a sophomore studying biology at the University of Arizona, you probably were. Facebook itself mandates truth in packaging by prohibiting users from impersonating "any person or entity."

If this rule is being followed, then Karl Marx, Anne Boleyn, and Kermit the Frog are alive and well. A small but substantial number of Facebook pages are "owned" by personages who aren't in our classes or paying tuition. As of December 2, 2006, there were seventeen Karl Marxes out there, four Anne Boleyns (including one with 306 Friends), and seven Kermit the Frogs. By comparison, Socrates of Athens only managed three doppelgangers.

Social Affiliations

Once you have an identity, you can begin forming social linkages. The primary way to build online relationships is by "Friending" someone else on Facebook. That person receives an old-fashioned email with the request, which must be accepted or declined on Facebook itself. The number of Friends people have on Facebook ranges from a handful to many hundred.

A second way of forming social affiliations is by joining a "Group." Groups may exist in the real world (such as the Podunk University Women's Volleyball Team) or only in the fertile imaginations of the beholders. Some of my favorites include "I Went to Private School But Liberal Guilt Makes Me Slightly Embarrassed to Admit It," "When I Was Your Age, Pluto Was a Planet," and "I Want to Be One of Erin's Super Friends!"

Online Interactions

Facebook à la late 2006 offered three methods for interacting with other people online. (Others have been added since then.) The first was the personal Message, which works like an email: It's sent asynchronously and arrives in your Facebook account. The second method was the Wall, a kind of electronic whiteboard. Occasionally, people post on their own Wall (sort of like an away message), but usually it's other people who write on yours.

And then there was Poke, a tool with no pre-attached function. When I "Poke" someone, that person receives a message on Facebook saying "Naomi Baron poked you." What does it mean to Poke someone or be Poked? Anything from "Hi" to "I'm trying to annoy you" to "I'm interested. Shall we get together?"

Privacy

Facebook continues to introduce an ever-larger range of Privacy Settings, through which users can prevent other Facebook denizens from accessing particular information on their Facebook pages. As with IM, you can block specific people outright from seeing anything about you on Facebook or you can give limited access to groups of individuals. For example, anyone might be able to find you on a general Facebook search but only undergraduate Friends at your own university could see your particular list of Friends.

Facebook Users

Who uses Facebook? Millions and millions of people. According to comScore Media Metrix, in 2006 Facebook was the seventh most popular site on the entire web with respect to total page views. According to Facebook, the typical user spends about twenty minutes per day on the site, and two-thirds of those with accounts log on at least once a day.[9]

Why Use Facebook?

Facebook was originally created as a cross between a tool for meeting new people and a platform for networking with people you already know. The "About Facebook" section of the Facebook site loftily declares:

> Facebook is a social utility that helps people better understand the world around them. Facebook develops technologies that facilitate the spread of information through social networks allowing people to share information online the same way they do in the real world.

The social networking part rings true. How Facebook "helps people better understand the world around them" is a more amorphous claim, to say the least.

What is certain about Facebook is the control it offers its users. In an interview with John Cassidy that appeared in the *New Yorker*, Zuckerberg suggested that

> The way you [increase the information supply] is by having people share as much information as they are comfortable with. The way you make people comfortable is by giving them control over exactly who can see what.... People want access to all the information around them, but they also want complete control over their own information. These two things are at odds with each other. Technologically, we could put all the information out there for everyone, but people wouldn't want that because they want to control their information.

Earlier in this book, we suggested that communication technologies are increasingly offering people the opportunity to manage their terms of interpersonal linguistic engagement. Facebook ratchets up this control by letting users orchestrate what information they are willing to share about themselves with one another.

Facebook also affords its users another kind of conversational clout by minimizing the amount of time you need to spend in communicating with someone else. We probably all know acquaintances who sometimes favor voicemail or email over face-to-face communication (even with close friends) to eliminate the time that otherwise would need to go into social pleasantries.

Later on we'll see that American college students sometimes prefer to send text messages on their mobile phones rather than make a voice call for precisely the same reason. Facebook users have discovered that the site offers them similar control. As Cassidy explains, "One of the reasons that the site is so popular is that it enables users to forgo the exertion that real relationships entail." Cassidy quotes a recent Harvard graduate who explains why Facebook is such an effective way of keeping up with former classmates: "It's a way of maintaining a friendship without having to make any effort whatsoever."[10]

Finally, the control theme surfaces in users' ability to stage their presentation of self on their Facebook pages in whichever way they choose. How you, quite literally, picture yourself is your call. So, too, is how you describe your tastes in music, reading, and politics. Chris Hughes, who works for Facebook, explains that the site is "not about changing who you are. It's about emphasizing different aspects of your personality."[11] Users clearly have gotten the message. In a study of Facebook that several of my students did in fall 2005, one of the women interviewed explained that a Facebook Profile "can be more an expression of who one wants to be rather than who one really is." Indeed, me "on my best day."

Who Owns Facebook?

I asked a roomful of college students in fall 2006, "Who owns Facebook?" They initially met my question with a quizzical moment of silence, followed by, "We do." In the eyes of its users, the site is much like pebbles on the beach, there for the taking. No one "owns" the Internet; Wikipedia is freely open for all to read and contribute to; Google doesn't charge users for its services. Facebook is a social networking tool, whose content is shaped by all those people with Facebook accounts. So they collectively own it, right?

Not quite.

In the world of social networking sites, the question of ownership has loomed large since Rupert Murdoch's News Corp. paid $580 million for MySpace in 2005. That figure paled before the $1.6 billion Google laid out for YouTube in October 2006. While Zuckerberg still retains control over Facebook (as of late 2007), he has long been rumored to be looking for a buyer willing to pay handsomely.

The issue of who owns Facebook came to the fore the first week of September 2006, when Facebook launched two new features: News Feeds and Mini-Feeds.[12] Both tools provided updates on changes that Friends had made to their own pages, enabling you to learn the moment you logged onto your Facebook account that "Mark added Britney Spears to his Favorites" or "your crush is single again."[13] But millions of Facebook users were aghast: What right did Facebook have to bombard them with information they hadn't asked for?

Within one day of the new features being added, tens of thousands of users had already signed online protest petitions. Students on campuses across the country spoke of little else. They contemplated boycotts. They felt personally betrayed.

Zuckerberg was, apparently, surprised at the outcry. Recognizing the need to respond, he posted a letter on Facebook with the title "Calm Down. Breathe. We Hear You." His tone was that of an owner. While admitting that "we know that many of you are not immediate fans" of the new feeds, he went on to remind Facebook users that this information was available anyway. All he had done was "nicely reorganize . . . and summarize . . . so people can learn about the people they care about."

Well, sort of. If you have several hundred Friends, each of whom makes one change to his or her site every few days, you are barraged with an enormous amount of "news" that is essentially spam. What's more, the Privacy Settings on which Facebook prided itself weren't properly in place with the original launch of the new feeds. Facebook users refused to "calm down." Zuckerberg found himself having to change tactics. His "Open Letter" of September 8 was written not in the voice of an owner but of an apologist: "We really messed this one up. . . . we did a bad job of explaining what the new features were and an even worse job giving you control of them."

Had Facebook actually been naïve in launching the new feeds? Perhaps not. During the period from December 2005 through March 2006, the number of unique visitors to Facebook was largely flat.[14] Since the site's profits come from advertising revenues, and ad sales are pegged to user visits, Facebook clearly needed a new strategy. Allowing businesses and other organizations (both within and outside of the United States) was one element. The new feeds were another. An independent blog reported in mid-October that the gamble paid off, with the number of Facebook page views increasing over 40 percent in the prior month.[15]

The September Surprise ended in a win-win situation. Facebook's business prospered, and its users came away feeling they still owned Facebook.

Why Is Facebook Interesting?

Not being a venture capitalist, my interest in Facebook lies in understanding the ways in which young people use the platform to construct and conduct social interaction with peers. I've been especially intrigued by the choices students make that enable them to control their conversation with others. While I had heard many anecdotes from my students and gained a little first-hand experience by creating my own Facebook account, I had no substantive, objective data.[16]

As we saw in the first half of this chapter, teenagers and young adults were already manipulating the terms of online linguistic engagement through IM long before Facebook arrived on the scene. Since both IM and Facebook can be used for presentation of self and for managing social relationships, I wondered how college students divvied up their communication on the two platforms. The way to find answers was to do a study.

• • • THE FACEBOOK STUDY

In the spring of 2006, my students and I drafted a paper-based questionnaire about Facebook. The questions fell into eight broad categories:

- Demographics (including how long you have used Facebook)
- General Usage Patterns (how often you use Facebook and for what functions)
- Information about Yourself (what is in your Profile)
- Social Affiliations (Friends and Groups)
- Online Interactions (Messages, the Wall, and Poke)
- Attitudes toward Access (should non-students, including parents, be able to access your site)
- Privacy (what, if any, Privacy Settings do you use)
- IM versus Facebook (on which platform do you use which functions)

We also provided respondents the opportunity to offer additional comments. Here is what we found.

Demographics

Our subjects were sixty undergraduates at American University (half male, half female), with an average age of twenty years, six months. Although the mean age for males and females was essentially the same, males had logged more experience with Facebook (nearly a year-and-a-half, on average) compared with females (closer to a year).

General Usage Patterns

Overall, 55 percent of our subjects logged on to Facebook at least daily, though females slightly outpaced males (60 percent versus 50 percent). Recall that Facebook-the-company reports that two-thirds of users log on at least once a day.

Facebook also indicates that the typical user spends about twenty minutes per day on the site. A study at Michigan State University found that undergraduates averaged between ten and thirty minutes daily.[17] Our students spent more time, averaging over forty minutes a day. Again, females were slightly heavier users, averaging almost forty-five minutes compared with males, who (on average) clocked in at forty.

We also wanted to know how many people's Facebook pages our students looked at in a twenty-four-hour period. The answer: about seven, with females looking at slightly more pages than males.

Information about Yourself

Our next questions focused on how often students tweaked the information they posted about themselves. To look at the Facebook news feeds, it might feel as if students are continually updating their pictures or personal data. In fact, they aren't. Fewer than 5 percent changed pictorial information at least several times a week, and only a smidge more (7 percent) altered personal data with that frequency. These data correspond reasonably closely to findings in a Facebook study that Matthew Vanden Boogart did in spring 2006, involving students on four university campuses. Although Vanden Boogart's survey used different time intervals from mine, he found that barely 2 percent of his subjects altered their profiles either hourly or weekly.[18]

Why, then, do those receiving Facebook news feeds perceive so much change? Because they have so many Friends.

Social Affiliations

The notion of "Friend" on social networking sites has generated a good deal of buzz. From the perspective of many youthful Facebook users, the more Friends the merrier. You may or may not actually know the people who ask to Friend you—or vice versa. But so what? Like sports trophies, what matters is that they pile up. Some of my students have confided that when they first contemplated joining Facebook, they were embarrassed they would have a paltry number of Friends linked to their name. At the extreme, among some groups of real-life friends, the volume of Facebook Friends you amass is a point of competition, with the numbers edging into the thousands.

How many friends does an average Facebook user have? There's considerable variation from clique to clique, campus to campus, age-group to age-group, and across time. Our sample yielded an overall average of 229 friends, with males substantially edging out females (males: 263; females: 195). Vanden Boogart's average came out at 272. A study done at HP Labs, which examined data from millions of Facebook accounts, calculated 178.[19] The Michigan State study found the number was between 150 and 200.

Given how casually most Facebook users accept Friends, we were curious to see how these online trophies measured up with students' circles of "real" friends. After asking how many Facebook Friends our subjects had, we followed up with this question: "Of these 'Friends,' how many are real friends (e.g., you might go to a movie or dinner together)?" Not surprisingly, these numbers plummeted. Females reported an average of 65 "real" friends, and males, an average of 78, yielding an overall average of 72 people with whom the students might actually socialize. The number 72 is a far cry from 229.

While amassing Facebook Friends can be an online sport, joining (or starting) Facebook Groups introduces elements of humor and sometimes audacity. Who wouldn't want to be a member of "Erin's Super Friends," or how many rebellious freshmen wouldn't be tempted to join Groups with names containing language their mothers forbade them from using? Yet unlike Facebook Friends, the numbers of Facebook Groups students join is more restrained. Our subjects averaged 15 Facebook Groups apiece (males: 17; females: 14). Vanden Boogart reported almost double that number—an average of 27, with the disparity between 15 and 27 probably reflecting differences in campus culture (or in sampling).

Again, we wanted to distinguish between simple online membership and real-world activity. When we asked students, "Of these Groups, how many conduct offline activities (e.g., social, academic, athletic)?" the average shrank to barely two.

Online Interactions

Moving from social affiliations to online interactions, we probed how students used the Message feature, the Wall, and Poke. At the time we collected our data, you had to be logged on to your Facebook account to know you had received a Message. When someone wrote on your Wall, you received an email (but then needed to log on to Facebook to see what had been written). If someone Poked you, you had to go on to Facebook to find out.

The Message function was used by students in our study, but not extensively. Only 7 percent of the subjects used it "several times a day" or "daily" (females more often than males), while 10 percent never used the feature at all. Among occasional users, females did somewhat more frequent messaging than males. Low use of the Message function is corroborated in the massive longitudinal study from HP Labs, which logged the 284 million Messages exchanged by 4.2 million Facebook users between February 2004 and March 2006. The verdict: Less than one message was sent (on average) per user per week.[20] Hardly an impressive number.

The Wall also received sparse traffic. Barely 8 percent made a Wall posting either "several times a day" or "daily" (the only people in this latter group being females). Fully 60 percent either "never" wrote on someone's Wall or did it "less often than several times a week." When students did post to a person's Wall, they overwhelmingly did so to say "Hi" or to be funny. Only about a quarter of the subjects wrote on other people's Walls either to give information or to continue a prior face-to-face or telephone conversation. The conversationalists were nearly twice as likely to be female (a third of the females used the Wall for a conversation versus 18 percent of the males). Among those who did Wall postings, their main reason for using the Wall (rather than sending an individual Message) was "so other people can also see what I'm saying." Only 4 out of the 60 subjects posted on their own Wall—and all were male.

Poke was another function with little usage. Only 10 percent used it "several times a day," "daily," or "several times a week," while almost 53 percent never used it at all. The HP study also found a paucity of Pokes: 79.6 million Pokes compared with 284 million Messages, which averages out to about one Poke per user every three weeks. Males in our study were slightly more likely to use Poke. A total of 40 percent of the females ever Poked other people online, while roughly 55 percent of males did so. Those who used Poke reported that their primary reason was to return a Poke they had received. Others Poked to be cute, to be annoying, or less frequently, to say "Hi" or flirt.

Attitudes toward Access

At the time we did our study (spring 2006), only people at educational institutions had access to Facebook accounts. That meant students, but also faculty and staff. We were curious whether our students perceived Facebook to be "their" site, with faculty or staff members who had accounts seeming like interlopers.

While only colleges and high schools had official right of entry to Facebook, we knew that outsiders were sometimes gaining access as well. The news media were reporting how potential employers, along with graduate and professional programs, were checking out Facebook pages as they screened candidates for jobs or admission.[21] Equally troubling to many students was the prospect that their parents might see their profiles and photos, many of which displayed diminished clothing and ample alcohol. Again, we asked students how they felt about such "outsider" access. All of these issues relate back to the question of ownership. Does Facebook "belong" to its student clientele (for whom the site was originally created) or are students essentially visitors who are allowed to use the tools but without any say over what those tools are or who else has the right to use them?

Nearly two-thirds of our student subjects had no objection to faculty and staff being allowed to join Facebook. Here are some of the comments we received:

Those indicating that faculty and staff should have Facebook access:
"They [faculty and staff] are part of the university community. It makes things more friendly."
"They are allowed to, but I don't think we should get in trouble for things depicted on facebook."

Those disagreeing that faculty and staff should have access:
"The purpose of the medium is for college students to connect with one another. Any introduction of figures of authority will only serve to limit the speech and expression of individuals. Those people [faculty and staff] do not have the right to access your phone conversations or personal records, so facebook should be no exception."
"Because it's a student website."

Actually, it's Zuckerberg's web site, and he can assert any authority he pleases over it. He has no interest in limiting speech or expression, since free speech does not interfere with getting users' eyeballs on ads.

Sentiment against allowing graduate or professional programs, or future employers, to access student Facebook profiles was much stronger: Three-quarters thought it was a bad idea. Their reasons for objecting?

"incriminating evidence, this is not how I would represent myself to an employer, this is for my friends."
"I don't want them to see my pics"

Among those who had no problem with such access:

"I don't have anything illegal"
"Because you choose what to portray yourself as on Facebook, so
they'll see more of who you actually are. It doesn't have to be about
incriminating photos."

The last comment returns us yet again to the theme of presentation of self: If prospective employers are going to size you up through Facebook, why not create a page that enhances your employment prospects?

Finally, there was the question about parents: "Would you want your parents to see your Facebook page?" The resounding answer was "No." In fact, only a lone female replied "Yes." In retrospect, I realize we should have provided a third option: "Don't Care." Reading some of the additional comments subjects wrote at the end of the questionnaire, it became clear that for at least some students, it didn't matter whether their parents saw their Profiles and photos. One male student wrote: "I am not dying for my parents to see the profile, but I would not care if they saw it." In the same vein, a female said that "While I wouldn't necessarily 'want' my parents to review my Facebook page, I certainly wouldn't mind showing them if they were interested." Perhaps the "Don't Care" group had innocuous Facebook pages. Or perhaps the students and their parents had already made their peace.

Privacy

Facebook offers a variety of Privacy Settings, enabling users to decide for themselves what cadre of people may view which portions of their Facebook postings. Over the months, the number of privacy options has increased. What's more, settings that used to be buried on the Facebook site and confusing to use have become more transparent. However, anecdotal evidence suggests that many loyal Facebook users remain unaware of these changes.

As of spring 2006, 56 percent of our subjects restricted either the people who could find them in a Facebook search or who could see their Profile. (Interestingly, a fifth didn't know Privacy Settings existed.) Roughly two-thirds of the students were willing to let anyone on Facebook find their basic page. However, students were somewhat more discerning about who could see their Profile: While three-quarters gave access to fellow undergraduates on our campus, only 64 percent gave graduate students the nod, half allowed viewing privileges to alumni, 43 percent admitted faculty, and 41 percent opened their Profile to staff.

We then asked whether students restricted access to personal Contact Information such as email address, AIM screen name, telephone number, or residential address. Their options: "My Friends," "My School," "Friends of Friends." A third said "Yes," they did restrict some access. Forty-five percent did nothing to limit visibility, and another 22 percent didn't know Privacy Settings existed for this kind of information.

Facebook enables users to personalize privacy yet further by blocking particular individuals from accessing the user's Facebook page. Only a handful of our 60 subjects—4 females—used this option in Facebook. By comparison, on their IM accounts, 26 of the students (17 females, 9 males) blocked specific individuals either in general or for a period of time.

The privacy issue is more complex than first meets the eye. Students who have the technological know-how to find their way easily around the web are often oblivious even to the availability of ready tools for precluding relative strangers from accessing birthdates, hometowns, and current addresses and phone numbers. A study at the University of North Carolina at Chapel Hill found that the 38 undergraduate and graduate (or professional) students surveyed were commonly "OK with strangers accessing my [social networking site's] profile."[22] If students presume they "own" Facebook, understandably they sense no more need to put their information under privacy wraps than they feel compelled to put their clothes away if they live in a dormitory single.

Many students in the United States remain curiously ambivalent about privacy. In a study of 189 undergraduates, a team at Carnegie Mellon University (CMU) established that when presented with a general survey about privacy attitudes, subjects displayed understandable apprehension about making information regarding their whereabouts available to strangers: "The highest concern was registered for [the statement] 'A stranger knew where you live and the location and schedule of the classes you take,'" with 46 percent choosing the maximum point in the Likert scale, "very worried."[23] However, a sizeable proportion of this same cohort provided their schedule of classes or address (or both) in their Facebook Profile. What's more, in the researchers' larger study (which also included 74 graduate students, along with 31 faculty and staff), 33 percent maintained it would be "impossible or quite difficult for individuals not affiliated with a university to access [the] Facebook network of that university." In the now-familiar words of one person interviewed, "Facebook is for the students."[24]

Yet unwanted contact from strangers and even identity theft are accidents waiting to happen. In an earlier paper, the authors reported that of the more than 4,000 students whose Facebook behavior they studied at Carnegie Mellon in June 2005, a minuscule "1.2 percent of users (18 female, 45 males)" changed the default visibility setting from "being searchable to everyone on the Facebook to only being searchable to CMU users."[25] One year later, *Newsweek* noted that only 17 percent of all Facebook members ever made changes in Privacy Settings.[26]

It may be human nature to believe nothing bad can happen on your home turf. Unfortunately for its student users, Facebook is not just a virtual extension of their campus lives but an increasingly public site and potentially in the cross-hairs of ne'er-do-wells spotting an easy target.

IM versus Facebook

Our final group of questions asked students to compare how they used instant messaging and Facebook. Since the two platforms have a number of overlapping functions (a place to develop a profile; an away message or Wall function for posting information available to a collection of people; a one-to-one messaging system), we wondered how students were balancing these instruments. Students in a Facebook focus group I ran (prior to finalizing the questionnaire) suggested that Facebook had largely replaced IM. Were they right?

Total Daily Usage

We began by comparing time on the clock: "On average, approximately how much time per day do you spend using IM and Facebook? Count time actually using IM or Facebook functions, not just having IM or Facebook open." Here is how the daily averages, in hours, stacked up:

	IM	Facebook
Females	2.0	.73
Males	2.35	.60
Total	2.18	.67

The difference is stark: far more time on IM than on Facebook. Males spent a little longer on IM than did females, while the ratio was reversed for Facebook.

These data need to be interpreted in context. The numbers represent student estimates, not precise measurements. In the case of Facebook, students typically log on to use the site and then log off when they're done. While the site is open, their attention may switch to other pursuits. Some of these activities entail true multitasking—like roommates who sit in the same room, IMing each other about the Facebook Profiles they are reading. Other multitasking is more sequential—check out someone's new photo album on Facebook, have a phone conversation, add a Group to your own Profile, load a new CD to iTunes, and only then close Facebook.

IM poses an even greater challenge for accurately reporting usage time. Many students keep IM open for hours (even days) on end. As we saw earlier, an enormous amount of multitasking goes on while college students have IM open—even while holding IM conversations. Without keeping a twenty-four-hour IM usage diary, it's hard to know how precise the two-hour figure might be. That said, it's probably fair to conclude that IM is used for roughly three times as long as Facebook.

Profiles

Our other two questions concerned profiles. One asked, "To get general information on someone you know, which profile (AIM or Facebook) do you look at first?" We also asked about updating profiles: "Do you update your AIM or Facebook profile more frequently?"

The replies clearly suggested that Facebook is the place to go for profiles. Four-fifths of the students turned first to Facebook profiles to get information on another person, and two-thirds updated their Facebook profile more frequently than their IM profile. The only gender difference came with regard to information-seeking. More than 85 percent of females chose Facebook over IM as their first source of information, while the proportion for males was roughly 77 percent.

• • •

Despite the meteoric rise of Facebook as an online social tool, IM has hardly faded into the electronic woodwork. Instead, students are now assigning different functions to the two platforms. Facebook has largely become the network on which you present yourselves to others, while IM retains its role as the basic form of online communication between individuals.

Support for this dichotomy comes from an interview-based study of Facebook that a group of my students did in fall 2005.[27] Interviewees commented that they "tended to send a Facebook message when they wanted to communicate something private, but not immediate. They saw the message function to be like email." Students also noted that "an IM conversation with someone you'd rather not speak with, or whom you do not know, is seen as more of an invasion of privacy than a message via the more relaxed Facebook system."

The Facebook Wall provided another alternative to an IM or a Facebook Message. Several students admitted using the Wall "when they wish to avoid talking to the person in question." One woman gave the example of happy birthday messages:

> She posted a happy birthday message on the wall of a casual friend rather than call or send an instant message because she did not want to start a conversation.... The wall is often used as a means of social avoidance; users try to keep up social ties without having to actually maintain them.

Looking at the overall relationship between Facebook and IM, the fall 2005 student research project summed it up well:

> The Facebook... allows the person to maintain a presence in an online community from a distance. Being someone's friend or joining a group carries no obligation or responsibility.... AIM takes on a more personal role, similar to that of a phone number or physical address... [While Facebook enables you to present yourself on your best day], AIM is a direct line to the user and their state of mind at any given time.

The Gender Question

Throughout our discussion of Facebook, we've noted small but consistent gender differences in how the site is used. For IM, we observed a number of divergences in the online conversations that male and female American college students construct. Of particular relevance here are the findings that females averaged longer IM conversations and used longer closing sequences before the final goodbye was said. We also found that the literature on gender and language repeatedly argues females are more prone than males to use both spoken and written language for social interaction.

The Facebook data are consonant with these trends. On Facebook, females were more likely to log on daily, spent a little more time each day on Facebook, sent slightly more Messages, were more likely than males to continue "real" conversations on the Wall, and were a little more likely to get information on people via Facebook Profiles rather than through IM. All this, despite the fact that males averaged longer experience with Facebook than females. Given our sample size, these differences were not statistically significant. Yet in raw numbers, the gender patterns were tantalizingly consistent.

We also observed that males in our study reported spending a little more time actively using IM than did females. While I don't have "pre-Facebook" comparative statistics on IM daily usage, by gender, it's possible that females are now moving some of their social functions from IM to Facebook. Alternatively, use of Facebook may simply be additive.

• • • SOCIAL NETWORKING FOR ALL?

Is social networking for everyone? Judging from the numbers, it's easy to think so. By June 2006—before taking the lid off membership requirements, Facebook had about 15 million visitors a month. These numbers pale before MySpace traffic: 21.8 million visitors in August 2005, which soared to 55.8 million by mid-2006.[28]

People aren't the only ones getting in on the act. Dogster.com was launched in 2004 and as of December 2006 had 229,372 members—each of which had its own web page, complete with a profile including a Biography, Pet Peeves, Favorite Toys, Favorite Food, Favorite Walk, Best Tricks, and even a list of Groups to which the pooch belonged. Catster.com follows the same format, clocking in at 97,402 members near the end of 2006.[29] Personally, I have not seen pigs fly nor dogs and cats logging on to computers. The heavy lifting is left to the pets' owners (or friends and family), who participate vicariously in the social network.

Not everyone, even of college age, is enamored with online social networking. Estimates from Facebook itself and from external studies suggest that between 80 to 90 percent of students on American campuses have Facebook accounts. In February 2006, the *Austin American-Statesman* reported that four-fifths of undergraduates at the University of Texas at Austin were on Facebook (along with a smattering of faculty and staff).[30] The numbers at Carnegie Mellon were slightly lower: 70 percent by late 2005.[31] Michigan State (as of April 2006): over 90 percent.[32] What about the rest?

Some students have no interest in joining in the first place. Others join but later drop out. Discussions with students who have opted out suggest their decisions are usually motivated by one of two factors. The first is time. Facebook can be a dangerous tool for procrastinating—when you should be cleaning your room, writing a paper, or studying for an exam. Vanden Boogart notes that of the 2,851 undergraduates he surveyed, a third either "Agreed" or "Strongly Agreed" with the statement "I feel addicted to Facebook."[33] As one student wrote, "Facebook, I hate you!"—for sapping so much of her attention.

The second reason students give for shunning Facebook is privacy. They have no desire to post a staged photograph, to give out their birth date or political affiliation to people they barely know, or to subject themselves to Pokes or to "news" flashes when one of their "Friends" updates a Profile.

But at least among college students, those who say no to online social networking have been in the minority. Much as a student in the IM away messages study said she owed it to her friends to let them know her whereabouts when she was not physically in front of her computer, a subject in my Facebook study wrote that "It allows people I know to see what I am up to." For substantial numbers of people, online communities enable them to see and be seen. And by controlling what information is posted, users help ensure they will be seen on their best day.

• • •

The last chapter looked at one-to-one communication, using IM. In this chapter, we have examined broader social networks (IM away messages and Facebook), in which users still restrict the size of their community. The next chapter expands the online social circle to the world of blogs, YouTube, and Wikipedia, where an individual's audience is potentially far more vast—and unknown.

6

● ● ● Having Your Say

Blogs and Beyond

Nothing, said Samuel Johnson, focuses the mind like a hanging. If history is to be believed, the prospect of hanging may also loosen the tongue. Nowhere is this truism better documented than in a West London spot once known as Tyburn.

As far back as the twelfth century, London criminals (or those judged to be such) were hanged in public from a tree named after the Tye Bourne, a brook that ran there. The place retained its grim function over time. By the seventeenth century, it was common for those about to die to confess their sins and ask for forgiveness, following a set formula.[1] The eighteenth century saw a surge of public hangings, mostly of the poor and disenfranchised. On their way to the gallows from Newgate Prison, the condemned were offered alcoholic spirits by local taverns along the route and as a result, commonly arrived at Tyburn rather drunk.[2] Felons were allowed to speak their minds to the crowd before meeting their fate.

The last public hanging at Tyburn was in 1783, whereupon executions were moved to the confines of Newgate. Nearly a century later, thanks to a series of reform movements and protests over political issues, an 1872 Act of Parliament set aside an area in the northeast corner of Hyde Park that could be used for public speaking. The legislation, formally known as The Royal Parks and Gardens Regulation Act, established the place we now know as Speakers' Corner near the very spot where the condemned of Tyburn had uttered their last testaments.

Karl Marx spoke there. So did Friedrich Engels, George Orwell, William Morris, Emmeline Pankhurst, and Marcus Garvey. For more than a century, Speakers' Corner in Hyde Park has epitomized free expression in the modern world. The procedure could not be simpler: You show up on a Sunday morning and hold forth. Over the decades, the concept of a "speakers' corner" has been adopted in many other parts of the world.[3]

Speakers' corners attract the passionately dedicated and the doggedly persistent, the articulate and the uneducated, those who are highly convincing and those who should, perhaps, be committed. Like modern-day bloggers, they have full jurisdiction over their words, and their would-be audiences are equally entitled to listen, heckle, or ignore the soapbox performance.

Of course, speakers' corners haven't been the only venues in which individuals can have their say. Two other pre-Internet platforms have been letters to the editors of newspapers and, more recently, talk radio. While control is not entirely in the hands of the would-be "speaker," both newspapers and talk shows offer access to a substantial audience.

What do speakers' corners, letters to the editor, and talk shows have to do with language online? By reminding ourselves of these highly popular precursors to blogs and other Internet-based platforms for individual expression, we discover how today's online tools are satisfying needs that have long been served in other ways.

• • • LETTERS TO THE EDITOR: HAVING YOUR SAY IN WRITING

Modern journalism is a product of the early 1600s, an outgrowth of handwritten newssheets, called *gazzette*, which first appeared in Venice in the mid-sixteenth century. These *gazzette* brought news of the rest of Europe to Italy. By 1632 the oldest newspaper we know of was printed in England. The first daily newspaper in London was established in 1702.[4]

Eighteenth-century England saw an explosion in the publication of newspapers and magazines. Daniel Defoe began a paper in 1704; the *Tatler* made its debut in 1709, followed by the *Spectator* in 1711. *Gentleman's Magazine*, which published until 1907, was launched in 1731. The growing periodic press was ripe for copy, including letters from readers.[5]

Early letters to the editor took a variety of forms. While a number were independently submitted, others were solicited by the newspapers themselves. Sometimes editors commissioned letters from correspondents in their employ or published letters written by friends. Such was presumably the case with the twenty-three letters that Enoch Cobb Wines, a Congregational minister, wrote to the editor of the *United States Gazette*, narrating Wines's journey from Philadelphia to Boston—reminiscent of contemporary travel blogs.

Like many of his day, Wines signed his name with a pseudonym: Peter Peregrine ('traveler'). Sometimes a pseudonym was chosen simply to preserve anonymity. In other instances, a name such as Publius was selected to imply that the piece was not personal opinion but the voice of the common man—a tactic adopted by Alexander Hamilton, James Madison, and John Jay in their eighty-five essays, initially published in New York newspapers, that became known as the *Federalist Papers*.

In the first century of letters to the editor, authors tended to have social standing. However, the platform was, in principle, open to anyone. Anonymity allowed "those either so lowly they ought not to presume to rise, or so high that they should not have sunk, to involve themselves in public debate."[6]

While many letters to the editor focused on political issues of the day, there was seemingly no limit to the topics that might show up. *Gentleman's Magazine*, for instance, carried letters dealing with theology, earthquakes, and ghosts, not to mention passages copied from manuscripts or rare books, complete with commentary.[7]

Literary and intellectual luminaries have often taken up the newspaper rostrum. Charles Lutwidge Dodgson (aka Lewis Carroll) wrote on a pair-combination method for scheduling players in round-robin sports events, such as lawn tennis, and on proportional representation in voting.[8] Arthur Conan Doyle's letters to the editor dealt with medicine, politics, sports, the law, foreign affairs, literature, military issues, religion, and in 1931, a series on building a Channel tunnel.[9] George Bernard Shaw (whose collected letters to the press numbered around 155) published on his objection to flogging as a form of punishment (legal in England until 1948) and used the newsprint pulpit to continue arguments with—or show support for—such men of letters and science as H. G. Wells, G. K. Chesterton, Julian Huxley, and J. B. S. Haldane. Yet in the end, Shaw was not certain that this time had been well spent. In the words of the editors of Shaw's public letters, "Shaw calculated that he lost at least four years of his life writing 'superfluous letters,' during which time he could as easily have written three good plays."[10]

By the early 1890s, there were more than two dozen daily papers in London. And they were big: over two feet high and eighteen inches wide. One reason letters to the editor proliferated was that given all that space, editors "could afford to be indulgent to readers who wished to communicate their thoughts and opinions to their fellows."[11]

A similar challenge faced the radio in its early days: Networks needed programming. What more inexpensive and audience-friendly format could one imagine than a conversation?

• • • TALK RADIO: "LAST BASTION OF FREE SPEECH"

The first talk-radio show was in the early 1920s, the honor probably going to a program about farming.[12] In the early years, hosts tended to monopolize the air waves. Take John J. Anthony, who held court on *The Goodwill Hour*. The program announcer opened each show with the declaration, "You have a friend and adviser in John J. Anthony" and "thousands are happier and more successful today because of John J. Anthony!" How did Anthony engender well-being in others? By paraphrasing—and replying to—letters and telephone calls from his listeners regarding life and the pursuit of happiness.[13]

Talk radio became a long-running success story. While some programs eventually became talk television (such as the *Larry King Show* morphing into *Larry King Live*), talk radio continues to draw audiences and participants because of its very portability. In the office, driving on the highway, cleaning house—a radio is nearly always convenient, and so, these days, is a phone.

How many talk-radio shows are there? As of 1996, there were officially 1,992.[14] A decade later, there were close to 4,000 talk-show hosts.[15]

How many listeners are out there? In 1996, 18 percent of adults in the United States listened to at least one call-in political radio talk show at least twice weekly.[16] By December 2002, 22 percent of Americans got their news from talk radio, up from 12 percent in 1995.[17] In their 2007 edition of *Radio Today*, Arbitron noted that for 2006, 47 million listeners tuned in to "news/ talk/information" shows each week, while more than 11 million listened to "talk/personality" stations.[18]

With the expansion of Internet radio (potentially increasing listenership) and the explosion in blogging (which might siphon off some of the audience), the actual number of listeners is a moving target. However, the staying-power of radio should not be underestimated. Arbitron reported that in 2006, 93 percent of Americans aged twelve and over were listening to the radio at least once a week.[19]

Statistics on listeners who call in are harder to estimate. A 1993 survey found that 11 percent of Americans had ever tried to call a talk-radio show, while 6 percent had made it onto the air.[20] Another study (this one from the late 1980s) reported that less than one half of 1 percent of listeners had ever called in.[21] In either event, there are a lot of lurkers out there.

I'm the Host, You're the Guest

The essential elements of talk radio are these: a host, one or more guests, and folks to ask questions—either members of a studio audience or people calling in on the phone (or these days, sending an email or text message). What makes a show successful isn't just the topics discussed, the expertise of the guests, or even the host's personality but also the skill with which the host handles questioners. From the perspective of the listening public, talk radio feels like what Jerry William, a Boston talk-show host, described as "the last bastion of freedom of speech for plain, ordinary citizens."[22] In fact, though, callers are generally at the mercy of the show's producers.

Speech is hardly free if your call is never taken on the air. Nearly all shows have screeners who find out what you want to talk about, determine whether you are inebriated or have an accent that might not be understood, judge if you will sound interesting or boring. If you make it through these hurdles and get the chance to go live, the host still holds all the cards. Hosts can, of course, hang up on callers, but they also have more ingenious forms of control. They might have the engineer turn down the volume on a caller who starts shouting. Automatic devices can block out the caller's voice if the host and caller attempt to speak at the same time. And the host can close the encounter with a "Thanks so much for your call," while the "ordinary citizen" continues on, unaware he or she is no longer on-air.

The balance of power between host and "citizen" varies from show to show. Some contemporary celebrity hosts seem to take perverse pleasure in being rude to callers they find offensive. Other hosts—especially on public radio stations—are models of decorum and work hard at ensuring that callers have their say.[23]

Why Is Talk Radio So Popular?

Talk radio is a bit like a Rorschach test, open to individual interpretation. For some, it's a form of entertainment, allowing them to take in the passing show. For others, it's a medium for education—assuming you can believe what you hear. A third function of talk shows is to perpetuate the idea of free speech. Similarly, talk television (in this case, in the UK), has been described as the embodiment of democratic ideals:

> On the television [talk show], the doctor is on an equal footing with
> any of his patients, just as the Archbishop of Canterbury is with a pop star
> and yet in real life that is not so. People are not equal in society, the lit-
> tle girl who works in the back of the shop is not on equal footing with
> a professor of Greek, and yet on the box, everyone is the same.[24]

Undoubtedly, talk-show devotees relish the opportunity to have their say (or to hear like-minded callers do the speaking for them). But an equally important reason for the popularity of talk radio is the companionship it provides. This quest for social connection applies to listeners and those calling in alike.

In the early 1970s, psychologists such as Stanley Milgram began suggesting that urban living encourages superficial relationships, resulting in the diminution of "psychic resources" such as status, love, and attention.[25] Talk shows were an obvious medium to help fill the interpersonal gap: They "serve...as a companion for lonely people, countering the growing isolation of many in modern society."[26]

A study of radio talk-show callers to a Philadelphia station in spring 1973 found that "the principal urge motivating people to call a 'talk station' is rooted in a highly personal need for communication—for contact with the outside world."[27] The ninety-seven callers who were interviewed seemed more isolated than the Philadelphia population at large (judging from the 1970 U.S. Census). Callers tended to be older, of lower socio-economic status, and more likely to be widowed. The demographic exception was housewives—younger, better educated—who called "during the morning and afternoon [as] victims of a temporary loneliness that accompanies their housework."[28]

Other research on talk radio and loneliness followed. A study in 1978 argued that "talk radio is more than a mere outlet for opinions. It is a medium for interpersonal communication."[29] A decade later, researchers compared those who called in to talk-radio programs with people who just listened. While both callers and noncallers used the medium as a form of companionship and for passing the time, those who called in

> found [face-to-face] communication less rewarding, avoided personal communication, were less mobile, felt talk radio was more important in their lives, [and] listened to more hours of talk radio each day.[30]

The authors concluded that talk radio provided those who called in "an accessible and nonthreatening alternative to interpersonal communication."[31] This theme is echoed in studies of talk radio from the 1990s:

> We may not know our neighbor next door; we may not want to; we may be afraid of the stranger or possible criminal on the street. But radio and TV talk shows have become welcome visitors that help us know what's going on and make sense of an increasingly dangerous, alienating world.[32]

Focusing on those who call in, journalist Peter Laufer suggests that

> Typically, callers to talk shows are seeking companionship. They are lonely, stuck at home, or stuck in traffic. They feel disenfranchised from society

and desire an opportunity to be heard; they are convinced they have something to say.[33]

Seeking companionship. Something to say. Keep these words in mind for our coming discussion of blogs.

Should You Believe What You Hear?

In 1993 the Times Mirror Center for the People and the Press found that the most important reason people listened to talk radio was to obtain information. This result, says Laufer, is an "unnerving finding," since "much misinformation is spread intentionally and unintentionally by talk radio hosts," and callers and their opinions are "thoroughly manipulated...by individual hosts and the talk radio system."[34]

Consider for a moment that over half of all talk-radio shows deal with general-interest topics, politics, or public affairs. In the eyes of their hosts, are these shows news or entertainment? If Laufer is correct, we shouldn't be surprised to hear talk-show host Rush Limbaugh proclaim in a 1991 interview that "I look upon my show as an entertainment forum for people" and that "the main purpose of a good call...on my show...[is] to make me look good."[35] Limbaugh is hardly alone in his sentiments. In 1994, a San Antonio talk-show host affirmed that members of his profession were "entertainers," not people "paid to formulate domestic and international policy."[36] Of course, news and entertainment need not be mutually exclusive: "more choices and shortened attention spans have led Americans to combine activities wherever possible. Now we want to be entertained while we are being informed."[37]

In 2005, the Pew Research Center for the People and the Press investigated the amount of trust the public had in different sources of news: television, newspapers, Internet news blogs, and talk-radio shows. Asked to rate each medium with regard to whether it "mostly report[s] the facts about recent news developments, or mostly give[s] their opinions about the news," those surveyed clearly saw news-oriented talk-radio shows as opinioned: 68 percent said "mostly opinion," while only 10 percent said "mostly facts."[38] Yet since truth ("facts") is generally elusive, we commonly make judgments on the basis of opinions with which we happen to agree—either because they are consonant with our prior belief structures or because we trust the people voicing them.

Laufer argues that the trust many listeners place in what they hear from anonymous callers (not to mention named hosts) reflects our need to connect with others in an anomic world:

people are seeking replacements for the loss of direct personal contact in modern society. . . . Few of us sit around a coffee shop chatting over the morning paper with our neighbors. We grab a coffee to go, drink it in the car during the boring commute, and tune in to the radio for company.[39]

News or entertainment? Accurate or not? Balanced reporting or shock-value hyperbole? Presciently, TV newsman Dan Rather addressed these questions in a letter to the *New York Times* on March 8, 1994, in which he pleaded that "If we keep blurring the distinctions and standards between news and entertainment, we're all going to have to pay. And I respectfully submit the price is too high."

When Guests Become Hosts: From Talk Radio to Blogs

Talk radio is built on a model in which the host is in a privileged position, controlling access to the broadcast equipment and deciding which callers will be permitted to have their say. Some talk shows are more democratic (or demagogic) than others, but callers always serve at the pleasure of the host.

What happens if you turn the tables? Much as desktop publishing put the power of the press into the hands of anyone with a computer and printer, the Internet turns each of us into a potential talk-show host. The medium, this time, is not spoken radio but the written blog.[40]

• • • PUTTING MARGE IN CHARGE: THE DYNAMICS OF BLOGGING

I should have guessed from the name of her blog that "Marge in Charge" was likely to hold strong opinions and not hesitate to express them. Not that I was a regular reader. I happened upon Marge in the course of research I was doing on how users of the Internet formulate an image of someone's identity based upon information revealed through web searches. Since the person I best knew was yours truly, I did a large-scale ego search on Google, Yahoo!, Lycos, and Alta Vista, drilling down several hundred hits on each. There, at number 155 on Google (as of May 8, 2004), I met Marge.

Marge's blog entry for March 20, 2003, dealt with a *New York Times* piece on IM away messages, in which I had been quoted regarding the study presented here in chapter 5. Apparently Marge was none too impressed with research on computer-mediated communication. Misreading the article, she began by lighting into me for teaching a semester-long offering on just away messages (which, of course, I had not). But since Marge was in charge, there was no one to correct her. She barreled on ahead:

> WHAT? You can take a class that analyzes away messages? Are you [expletive deleted] kidding me? . . . *Away messages?* I'm going to do an independent study on e-mail signatures. I want a B.A. in answering machine anthropology . . . [more expletives] . . . I'm transferring to American so I can get a Ph.D. in AIM.

For readers who didn't know me (or the legitimate research field of CMC), I probably came across as either socially hip or an academic slacker. I took only small consolation in my observation that more than a year after her tirade, the number of comments posted to her blog entry was zero. My pride was hurt, but I felt reasonably assured that my reputation (and that of my university) remained intact.

Dan Rather was not so lucky. On September 8, 2004, *CBS 60 Minutes Wednesday* aired a piece Rather had done about George W. Bush's experience in the Texas Air National Guard. The story suggested Bush had used family influence to evade the draft (this was the Vietnam Era), and that he didn't actually fulfill his National Guard obligations. At 11:59 p.m. that night, the first negative reaction was posted on the highly conservative blog site FreeRepublic.com. A cascade of other bloggers quickly picked up the story—PowerlineBlog.com, DrudgeReport.com, and the newly minted RatherBiased.com. Their argument: The *60 Minutes Wednesday* account was based on forged documents. CBS News conducted an internal investigation, which added more fuel to the blogging fire. Both CBS and Rather eventually apologized for running the story without ample certainty of their facts, but bloggers were calling for Rather's head. On November 23, Rather announced he would end his long career as anchor of *CBS Evening News*, which he did on March 9, 2005—twenty-four years to the day after assuming the post. (In September 2007, Rather filed a $70 million lawsuit against CBS, arguing that he had been made a scapegoat.)

The incident is reminiscent of a trial-by-talk-radio that occurred a decade earlier. The 1993 nomination of Zoe Baird to be U.S. attorney general was defeated not by the Clinton administration, the Congress, or even the press. Instead, it was largely talk-radio callers voicing strong disapproval of the terms under which Baird had hired a nanny that led to the nomination being withdrawn.[41]

With the *60 Minutes* case, the blogosphere meted out one of its first instances of "justice" on the electronic frontier. Rather was hardly the first journalist not to have gotten a story exactly right, but this time his judge and jury were a small, vocal, politically opposed cluster of bloggers.

When many people talk about blogs, they have in mind the online writings of a relatively restricted number of individuals, most of whom deal with political issues or at least current events. These are the so-called A-list bloggers, the ones whose readership surpasses that of many mid-sized newspapers. But when it comes to finding an audience on the Internet, most bloggers have more in common with Marge.

What are Blogs, and Where Did They Come From?

The humble beginnings of blogs trace back to the late 1990s, when a handful of web denizens began compiling lists of the URLs for online pages that the creator of the list found to be interesting. Since these listings were literally logs of web locations, Jorn Barger's term "web log," coined in 1997, made eminent sense. Early web logs included Dave Winer's Scripting News, Cameron Barrett's CamWorld, and Barger's Robot Wisdom, though some purists date the first web log back to 1991, when Tim Berners-Lee, father of the World Wide Web, created "What's New?" pages that linked to other web sites.

Many early web logs consisted of just headlines accompanied by links to pages bearing the actual stories. Other web logs offered brief news summaries or discussions of contemporary topics, alongside the links. One rule of the game was that the logs needed to be updated frequently.

Web logs soon gained in popularity. With increased visibility, the phrase "web logs" morphed into *blogs*, much as "God's blood" became shortened to *'sblood* in Shakespeare's day. Free blogging tools made it easy to join the blogging revolution. Ever-more powerful search engines began bringing up hits on the blog postings of obscure bloggers, and hence my encounter with Marge in Charge.

Several people who write about blogs have noted parallels between blogging and talk radio. The analogy was nicely articulated by Bonnie Nardi and her colleagues:

> Just as with radio, the blogger can broadcast messages of their choosing,
> without interruption. Limited feedback analogous to listener call-in
> on a radio station is possible with comments on blog posts. The comments remain "subservient" to the main communication in the posts,
> just as a talk show host or deejay dominates listeners.[42]

In fact, one blogging software platform is called Radio UserLand. Other antecedents to blogs have been noted in the literature as well. Speakers' Corner is sometimes invoked, as are letters to the editor and traditional handwritten diaries.[43]

Who Blogs, and About What?

As with talk radio, there are blog writers (compare "hosts" and "callers") and blog readers ("listeners"). How many are there of each?

In July 2006, the Pew Internet & American Life Project reported that about 12 million American adults blog (8 percent of Internet users), while about 57 million American adults (39 percent of Internet users) read blogs.[44] Other research found that 9 percent of people surveyed said they read a political blog "almost every day."[45]

Involvement with blogging varies across cultural milieu. In the UK, only 2 percent of Internet users wrote blogs in early 2006 (a quarter of the U.S. figure).[46] What is more,

> only 13% of those surveyed in the UK had read an individual's blog in the preceding week, compared with 40% in the US, 25% in France and 12% in Denmark. 12% of UK readers had read a newspaper blog in that week, compared with 24% in the US, 10% in France and 9% in Denmark.[47]

Getting precise tallies on blogging is difficult because the medium invites transience. I've created four or five blogs for my classes over the past few years. When the course is over, the blog remains, floating like space junk—but in cyberspace. Am I a blogger? Not really, but I have several blogs in my name. Do I read blogs? Sometimes, but mostly when they turn up in web searches. If a survey asked me whether I write or read blogs, and how many blogs I have, I could at best confound the data.

The scope of blogs has expanded dramatically over the past decade, making blogging more of a multifaceted tool than a specific type of Internet platform. Most blogging services offer a range of options the blogger controls: Who may read the blog in the first place? (Is the blog password-protected? Does it have an unlisted URL?) Does the blog take comments? In these regards, access controls on blogs are akin to privacy settings on Facebook. In fact, the similarity between media goes further. On some blogging sites, the opening page contains a profile of the blogger that looks amazingly similar to what you find on Facebook or MySpace.

The control issue becomes particularly nuanced with individual blogs primarily intended for a recognized circle of friends. On these kinds of blogs, the average number of comments posted by other people approaches zero,[48] which is probably all for the better. As Nardi and her group found in their interviews with adult bloggers, "Bloggers wanted readers but they did not necessarily want to hear a lot from those readers." Why? Because "many bloggers liked that they could be less responsive with blogging than they could be in email, instant messaging, phone, or face to face communication. They seemed to be holding their readers at arm's length."[49]

• • •

In principle, all blogs share four basic features. They're predominantly text-based (though graphic supplements are becoming increasingly common). The entries appear in reverse chronological order (that is, most recent first), with an archive kept of earlier postings. Blogs are frequently updated, and blogs contain links to other web sites. In practice, only the first two criteria are consistently found in today's blogosphere. As the medium attracts ever-larger numbers of users, bloggers are deciding for themselves how often they wish to post and whether they care to invite you (via links) to other URLs. A study of more than 5,000 blogs (done by Susan Herring and her colleagues at Indiana University) found that 42 percent were not linked in any way to other blogs.[50]

Over time, blogs have evolved from strictly current events sites to arenas for more varied self-expression. Herring and her students distinguish between three basic genres of blogs.[51] The first is topical (or "filter") blogs, exemplified by the news-based and political blogs that have figured prominently in the popular media's discussion of blogging. Examples include Matt Drudge's The Drudge Report or Markos Moulitsas Zúniga's Daily Kos. The second category is personal journals or diaries, following the earlier lead of web diary sites such as LiveJournal.[52]

This genre also might include travel blogs intended for readers back home, open blogs from soldiers in Iraq, or replacements for the annual Christmas letter. Finally, Herring talks about so-called knowledge blogs, in which individuals share their expertise. Examples include John Baez's This Week's Finds (on issues relating to mathematical physics) or Lawrence Lessig's Lessig Blog (which discusses legal and political issues involving copyright, especially in an online world).[53] Other researchers have identified additional blogging genres, including support groups.[54] Another category we've already mentioned is blogs created for academic purposes.

When you probe beneath the typologies to actual statistical sampling, it turns out that the blogs getting the least press are actually the most prevalent. Of the 203 blogs that Herring and her students analyzed (randomly selected through the blog-tracking web site blo.gs), more than 70 percent were of the personal-journal variety. Another study using random data from

Jeff Stahler: © Columbus Dispatch/Dist. By Newspaper
Enterprise Association, Inc.

blo.gs examined the demographics of bloggers: How many were males versus
females, and of what age? Using two samples drawn in 2003, researchers at
Indiana University report that while the overall numbers of male and female
bloggers was nearly the same, there were more female teenage bloggers but
more adult male bloggers. Among adults, topical blogs—which capture media
attention–were overwhelmingly written by males.[55]

Not surprisingly, the style of writing found across blogs varies with respect
to genre. Personal journal blogs are more likely to use language that some
researchers say exemplifies "female" writing style, while topical/filter blogs
contain more "male" stylistic features. Interestingly, this generalization holds
true regardless of the author's actual gender.[56]

Why Blog?

People choose to write blogs for an array of reasons. According to the Pew
Internet & American Life Project, adults' top four motivations for blogging in
2006 were

- to express themselves creatively
- to document their personal experiences or share them with others
- to stay in touch with friends and family
- to share practical knowledge or skills with others[57]

The first three reasons suggest that a lot of personal journal writing is taking place. The final category—sharing knowledge or skills—sounds like knowledge blogs. Looking ahead in this chapter, this last group of bloggers seems ripe for participating in Wikipedia.

Bloggers in the Nardi study also shed light on why people blog. Subjects identified five main reasons:

- update others on activities and whereabouts
- express opinions to influence others
- seek others' opinions
- "think by writing"
- release emotional tension[58]

A number of these motivations also appear on the Pew list (whose fifth through eighth runners-up included "to motivate other people to action" and "to influence the way other people think").

Let's dwell for a moment on the function of releasing emotional tension. Several people interviewed in the Nardi project talked about using blogs to "let off steam" or needing to "get it out there."[59] Elsewhere, teenagers report similar motivations. In the words of one sixteen-year-old who kept a web diary, "When there were days when I just needed to rant, it felt good."[60] Why blog in these circumstances rather than talk with people face-to-face, on the phone, or through IM? As another teenager commented, "blogs let writers interact while avoiding the emotional risks of one-to-one conversation."[61] Echoing this sentiment are adult bloggers who sometimes wanted an audience before whom to bare their souls but "desired to keep that audience at arm's length."[62]

Why else blog in place of personal conversations? Beyond some obvious reasons (reaching many people simultaneously, keeping a log of earlier posts, including lots of photographs), another motivation is not intruding upon friends. Blogs are a "pull" technology (like web sites you find on your own) rather than a "push" medium (such as email or IM, which shows up uninvited on your electronic doorstep).[63] Readers may choose to open a blog at their leisure—or ignore it altogether, while an email or IM is more in your face.

Finally, some people blog for money. Ana Marie Cox, aka Wonkette, began her blog through a paid arrangement with Nick Denton (publisher of

Gawker Media), who generated income by selling ads on the site. Advertising is now common on A-list topical/filter blogs, which can generate over a million dollars in revenue a year. At the micro level, ad revenue sometimes finds its way into the pockets of small-time bloggers, who are paid a few dollars to post complimentary words about products or services, though other bloggers post such encomia for free.[64]

• • •

In discussing talk radio, we identified four reasons that people listen to—or call in to—such shows. All four apply to blogs as well. Reading (and writing) blogs is a form of entertainment, though sometimes the entertainment verges on personal fantasy. The executive director of Wiredsafety.org notes that teenage bloggers have been known to portray themselves as engaging in risqué behavior, even when such is not the case—like the "girls who had blogged about weekends of drinking and debauchery, while in reality they were coloring with their younger siblings or watching old movies with Grandma."[65] Fact or fancy, these teens were controlling the medium, much like the college student who posted the away message about her enviable evening out on the town when she was actually home in front of the TV.

Reading blogs can be educational—again assuming you believe what you read. In their 2005 study, the Pew Research Center for the People and the Press found that only 20 percent of Americans judged news blogs to be trustworthy (that is, "mostly facts"). This was the same study that reported only 10 percent of Americans judged the news they heard on talk-radio shows to be "mostly facts."[66]

For bloggers, the medium constitutes an important platform for free speech. As with Speakers' Corner, potential readers "can choose to listen or walk away."[67] No one demands your credentials before you take the floor. But bloggers' freedom extends even further to controlling audience access and audience input (a bit like the host, rather than the caller on talk radio).

The fourth role of talk radio—companionship—is equally important for blogs. Many people read and write blogs out of loneliness and isolation. A perfect example is Mommy Blogs, which became popular in the United States in the mid-2000s. Imagine yourself a stay-at-home mom with three children under the age of six. For years, parenting and women's magazines have run stories about educated, accomplished mothers feeling they are losing their sanity, between the tantrums, diapers, and repetitive monosyllabic conversations. Enter Mommy Blogs, where you can blow off steam, remember that you still know how to craft an English sentence, upload pictures of your little ones, and take solace—by reading other Mommy Blogs—that you're not alone.

When we talk about blogs, we generally focus on the written text they contain. But modern blogging software often welcomes visual display. What if you cut out the text entirely and only upload graphics? If the images are still photographs, you enter the world of photo-sharing, made popular by such services as Flickr. If you're into movies, welcome to YouTube.

• • • NUMA NUMA ANYONE? WELCOME TO YOUTUBE

When sons and daughters go off to college, their parents often feel left out. Yes, there are phone calls home, but from Mom and Dad's perspective, most of the action is happening in unknown settings, with unidentified friends and acquaintances. My son's first kendo tournament was such an event. He was somewhere outside Detroit, *shinai* (bamboo sword) in hand, encased in his *do* (chest protector) and *men* (face mask), doing stylized battle with an adversary from I knew not where. What I would have given to be a fly on the wall of the gymnasium.

The price of admission turned out to be free. Teammates of the worthy opponent had made a video of the round and, some time later, posted it to YouTube. Cascading one online technology after another, the opponent matched my son's university with his last name—prominently displayed, as for all kendoists, on his *zekken* (name plate)—and proceeded to Friend him on Facebook.

The People's MTV

Who besides a parent—and members of the two kendo teams—might be interested in viewing other people's amateur videos? Back in the 1950s and '60s, guests used to cringe at the prospect of an evening watching the home movies of friends who had invited them to dinner. Today, however, our capacity for viewing badly shot images of other people's antics seems to be endless. Gary Brolsma, a teenager from New Jersey, achieved near-instant fame in late 2004 when he uploaded to newgrounds.com, one of the early Internet video-sharing platforms, a rendition of himself lip-synching "Numa Numa."[68]

Some background: In 2003, the Moldovan band named O-Zone recorded a song called "Dragostea din tei," which became a best-selling hit in Europe. The words *numa numa* appear in the chorus of the song, meaning something along the lines of "you won't, you won't take me." Other versions of the piece appeared in Europe, but for our story, the important rendition was a parody done in Japan, which substituted Japanese lyrics in place of the original Romanian by using words that sounded most similar in Japanese.

It was this Japanese version that Brolsma took as the basis for his own video. Brolsma's performance can now be found on many video-sharing platforms (including Google Video and YouTube) and has been played millions of times. That number doesn't include audiences of the countless other versions—made by Austrians, Brazilians, Spaniards, Belgians, Russians, Chinese, French, Finns, and more Americans than most of us have the patience to watch. In 2005 and 2006, viewing home-brewed versions of "Numa Numa"—often dancing along—was a popular high school and college party pastime.

Andy Warhol spoke of everyone getting fifteen minutes of fame. Video sites such as albinoblacksheep, newgrounds, Google Video, and YouTube afford us the possibility of extending that allotment by packaging ourselves online and hoping for an audience. Sometimes these video forums are simply convenient places for posting material intended for a small circle of friends. (Think of the kendo video, but also recall IM away messages or the Wall in Facebook.) If others happen to see your posting, no harm done. More often—as with Speakers' Corner, letters to the editor, talk radio, and open blogs—people doing the uploading have their eye on a larger audience. Like the proverbial Hollywood starlet, they just might be "discovered."

In our post-Napsterian universe, hundreds of upstart musical groups use these video platforms to gain free airings of their songs. For these users, it's important that the audience remembers their name. Other contributors to YouTube and the like are less concerned that viewers know their personal identity than that the song or vignette went up.

Under what circumstances do we care if our name is attached to our creative productions? The notion of copyright protection—for writings, music, photographs—is quite modern. Copyright as we understand it today did not exist for written works until the eighteenth century.[69] Not until 1831 was music deemed intellectual property by American law; photography was not included under U.S. copyright until 1865.[70]

Copyright law ensures not only that the writer, musician, or photographer is solely privileged to profit financially from the work but also that no one may tamper with the production without the author's (musician's, photographer's) permission. Why, then, in the twenty-first century would thousands of people work on an open-ended writing project for which there is no remuneration and in which only their words, not their names, are known? Such is the enigma of Wikipedia.

• • • THE PEOPLE'S ENCYCLOPEDIA: WIKIPEDIA

Do you want to know when Guglielmo Marconi won the Nobel Prize in Physics? What about the number of plays that Shakespeare wrote, the contents of Immanuel Kant's *Critique of Pure Reason*, or the consequences of global warming? Traditionally, these were questions that sent you to the library. Today, the online wonder known as Wikipedia is typically the destination of first resort.

Wikipedia allows anyone to have his or her say. And although names don't appear on the main pages of the articles themselves (in part because entries generally reflect many contributors' edits), an active social networking site functions below the surface.

The story behind this new twist on Speakers' Corner (and its descendents) begins in Hawaii.

The Wiki Wiki Line

Need to change terminals at Honolulu International Airport? Then take the free shuttle known as Wiki Wiki. Want to work collaboratively on a document, where everyone may add or edit text? Do so with a wiki—thanks to Ward Cunningham.

In 1995 Cunningham adopted the word *wiki*—meaning 'fast' or 'quick' in Hawaiian—to name a new tool for doing jointly authored writing, using the Internet. With a wiki, a number of individuals are able to write new content in a document but also to edit what others have written. Instead of Person 1 needing to draft a document, then ship it off to Person 2, who does editing and then forwards the revised text to Person 3, everyone can participate at essentially the same time. Wikis have become common in offices and organizations in which many contributors have their hand in a project. Free online software tools make for highly accessible systems, often used by small groups of colleagues who know—and trust—one another.[71]

But the same wiki tool can also be used by total strangers. The best-known example is Wikipedia.

Jimmy Wales Meets Denis Diderot

What are the goals of an encyclopedia? The most obvious is to gather together knowledge, presumably as objectively and accurately as possible, but sometimes with an underlying philosophical agenda. Another motivation is to make this knowledge available to your intended audience. The scope of that audience may differ from one encyclopedia to the next—some are written for the highly educated, while others are designed for children or the masses. A third purpose, less often articulated, is to present that knowledge through articles that model excellence in prose.[72]

The first important encyclopedia in English (though named in Latin) was John Harris's *Lexicon Technicum*, published in 1704. In 1728 Ephraim Chambers's famous *Cyclopaedia, or Universal Dictionary of Arts and Sciences* appeared, a book that would serve as a model for many subsequent works. In fact, it was the *Cyclopaedia* that provided the impetus for one of the most famous encyclopedias of all time: the eighteenth-century Enlightenment project known as *Encyclopédie: Dictionnaire Raisonné des Sciences, des Arts et des Métiers* ("*Encyclopedia: A Reasoned Dictionary of the Sciences, Arts, and Trades*").[73]

Initially, the plan had been to do a two-volume translation (into French) and expansion of Chambers's earlier publication. However, with the hiring of Denis Diderot and Jean d'Alembert to direct the project, the work turned into an Enlightenment manifesto, arguing through example the importance of scientific and rational thought, while at the same time demonstrating religious tolerance and challenging sectarian dogma. Thinking back to our three potential goals of any encyclopedia, the primary aim of the *Encyclopédie* was the embodiment of knowledge, but with a firm eye toward the modern, rational, and practical.

For whose benefit was this knowledge amassed? The public at large. All manner of knowledge was addressed: theology but also mythology, geography but also the making of Gobelin tapestries; natural law but also the nature of werewolves. Publication began in 1751. By the time the project was essentially done in 1772 (the work had been issued piecemeal), there were twenty-eight volumes, containing over 70,000 entries.

The *Encyclopédie* had more than 140 contributors, including the likes of Diderot and d'Alembert themselves, Jean-Jacques Rousseau, Voltaire, Marquis de Condorcet, Baron de Montesquieu, men of wealth, and members of the clergy, alongside people of more modest stature. Much of the *Encyclopédie*'s reputation derived from the caliber of the authors (and the writing) that appeared in its volumes.

Across the English Channel, a group of Scotsmen were planning their own encyclopedic work, largely as a conservative reaction to the Enlightenment (which, for some, meant godless) *Encyclopédie*.[74] The first edition of the *Encyclopedia Britannica* was completed in 1771, with much of the text written by its editor, William Smellie, who drew upon sources ranging from Francis Bacon's essays to Chambers's *Cyclopaedia*, Hume's and Locke's essays, the writings of Voltaire, and a spate of magazines and newspapers.

In the English-speaking world, the *Britannica* came to symbolize authority. Some of that authority derived from its writers, who, in the early nineteenth-century editions, included William Hazlitt, John Stuart Mill, Thomas Malthus, David Ricardo, and Walter Scott. The famous eleventh edition (which appeared in 1911) is known not only for the authors who contributed—including Charles Swinburne, T. H. Huxley, G. K. Chesterton, Ernest Rutherford, and Bertrand Russell—but also for the elegance of its writing. More recent editions have sometimes attracted expert contributors (such as the economist Milton Friedman, the cardiac surgeon Michael DeBakey, and the astronomer Carl Sagan).

Writing style has not been a modern hallmark of contemporary English-language encyclopedias. In the trenchant words of Charles Van Doren, a senior editor at *Britannica* in the 1960s,

> the tone of American encyclopedias is often fiercely inhuman. It appears to be the wish of some contributors to write about living institutions as if they were pickled frogs, outstretched upon a dissecting board.[75]

Robert McHenry, a former editor in chief of *Britannica*, wistfully recalls the poetic beauty—and strongly individual perspective—of such entries in the eleventh edition as Swinburne's biography of John Keats. McHenry candidly acknowledges that more recent editions have "a certain flatness, which is then represented by the likes of me as a virtue, the 'encyclopedic voice.'"[76]

Why should writing style be an issue in an encyclopedia? To the extent that a work is widely read—think of the eloquent and much-thumbed King James Bible—its style becomes a model to follow in our own composition.

Admittedly, accuracy is usually a more pressing concern. In his 1964 book *The Myth of the Britannica*, Harvey Einbinder argued that (at least in the mid-twentieth century), *Britannica* was laced with out-of-date information—duplicated from earlier editions and betraying ignorance of contemporary research, along with instances of out-and-out mistakes. In most cases, said Einbinder, the publisher was trying to produce volumes on the cheap rather than insisting upon the highest levels of modern scholarship.

Enter computers. With the new technology came the possibility of having knowledge at your fingertips—at least in principle. Initially, encyclopedic knowledge accessible by computer came in abridged formats on discs or then CDs, such as Compton's *MultiMedia Encyclopedia* (1989) and Microsoft's *Encarta* (1993). Then *Britannica* issued its complete encyclopedia on CDs and subsequently online—but with a comparatively hefty price. Knowledge, yes, and in principle for all, but beyond the fiscal reach of most.

Meanwhile, a seemingly unrelated set of developments was taking place in the computing world that would help redefine our conception of an encyclopedia. The roots of these changes go back to the days before the Internet, even before the personal computer, with the creation of a computer operating system named UNIX.

Around 1970, Ken Thompson and Dennis Ritchie, of AT&T's Bell Labs, wrote UNIX as an operating system that could run on a variety of computer platforms. (Previously, individual operating systems were prepared for machines built by different vendors.) AT&T made nearly all of the source code for the program available to universities, which promptly began to innovate. The best-known spin-off was BSD (Berkeley Software Distribution) UNIX, coming out of the University of California, Berkeley. BSD UNIX circulated among many institutions, where programmers modified the system to improve its functionality—not exactly a wiki but a collective enterprise serving a common goal.

In 1984 Richard Stallman from MIT launched a project known as GNU, under the aegis of his Free Software Foundation. GNU aimed to create a free version of UNIX that individuals could use, modify, and redistribute.[77] The dream became a reality in the early 1990s, when Linus Torvalds, a Finnish software engineer, wrote the kernel of a new operating system (called Linux), which he coupled with material from the Free Software Foundation, BSD UNIX, and additional software from MIT. The result was a genuinely open-source version of UNIX, for which the source code was freely available to be used, viewed, modified, and redistributed by anyone.

That same spirit of sharing was at the heart of development of the WELL (the early social-networking bulletin board created in San Francisco by Steward Brand in the mid-1980s) and evident in the unpaid contributions of programmers to UNIX and Linux.[78] It was also central to Project Gutenberg, an undertaking begun by Michael Hart in 1971, which now posts to the Internet (again, without cost) electronic versions of books and other writings that are no longer protected by copyright. Google Book Search goes one step forward (and another back), offering text from a vast collection of current titles, but only in small snippets at a time.[79] Out of this culture of sharing, coupled with Ward Cunningham's wiki, Wikipedia was born.

Wikipedia is the outgrowth of an earlier free online encyclopedia project, Nupedia, founded in 2000 by Jimmy Wales, with Larry Sanger as editor in chief. Nupedia articles were written by experts in the relevant fields and subject to extensive peer review. The labor of authors and reviewers was volunteered, much as happens in the traditional academic world. The project's goal was to create a work that rivaled in quality traditional commercial encyclopedias.

A year later, Wales and Sanger began Wikipedia, another free online encyclopedia, but this time based on a different principle. Using a wiki platform, anyone (not just experts) was invited to contribute articles as well as to edit those posted by others. Initially, Wikipedia was envisioned as a conduit for generating contributions that would then be peer reviewed and eventually incorporated into Nupedia. The problem Nupedia was encountering is one all-too familiar to authors and journal editors: Crafting accurate, insightful, and artistic prose is arduous, time-consuming work. Three years after its inception, Nupedia had vetted only two dozen articles—hardly encyclopedic. In September 2003, Nupedia ceased to exist, its contents being assimilated into Wikipedia.[80]

In short order, Wikipedia became an Internet phenomenon. As of May 2007, it had sites in 249 languages. The largest—English—boasted more than 1.8 million entries, putting the likes of *Britannica* to shame. The project is "an effort to create and distribute a multi-lingual free encyclopedia of the highest possible quality to every single person on the planet in their own language."[81] Since aligning itself in 2005 with such search-engine giants as Google and Yahoo!, which provide server space and bandwidth, Wikipedia has gained not only in breadth but brawn. Type a query into these search engines and, with surprising frequency, the first hit you get is a Wikipedia article.

As with a blog, anyone with access to the Internet can be a Wikipedia author, having his or her say. Unlike blogs, in which you may say whatever you please (libel and decency issues notwithstanding), the goal of Wikipedia is to provide objective knowledge. Readers who believe they have a better handle on such knowledge than you are free to alter your posting, a feature that sometimes leads to serial doings and undoings. The end result is that the contents of an entry may change at a moment's notice. As the Greek philosopher Heraclitus might have said, you may or may not be able to step into the same Wikipedia entry twice. However, the history of all edits (including those that others have trumped) is maintained.

Authorship Wikipedia Style

Wikipedia has a handful of fundamental "pillars" that define the encyclopedia's character and composition.[82] The first: No original research is allowed. The second pillar is that writing must be done from a neutral point of view. If you believe, as do most contemporary philosophers of science, that all observations are theory-laden, such a perspective may be unattainable. But the spirit in which the guideline is defined is understandable. If you're writing about the solar system and assert that UFOs definitely exist, you're not writing from a neutral point of view. If there is a controversy, then (says Wikipedia) you should present all sides of the story.

As Wikipedia has grown, it has become increasingly uniform in style. Through the urgings (sometimes not so gentle) of fellow writers, contributors progressively prepare entries using a standardized format, include references, and embed links to other Wikipedia articles. A frequent request is for someone to "wikify" a piece. (Recall McHenry's allusion to the contemporary "encyclopedic voice.")

As of September 2006, there were 75,761 "active Wikipedians" worldwide, with 43,001 of them contributing to the English edition.[83] Of the English-language contributors, more than 4,000 were especially industrious, making over 100 edits a month.[84]

What most users of Wikipedia don't see when they read an article is the world beneath the surface of the text. It's this world that raises Wikipedia from being "only" a free online encyclopedia (the user perspective) to an active social community of writers, editors, and sometimes camp followers. For contributors, the result is a cross between a cyber-exchange of letters to the editor, a listserv, a blog, a massively multiplayer game, and a social networking site. Although the elements continue to evolve, here are some examples of the community foundations as of January 2007.[85]

Editing Hierarchies

Anyone may write and edit articles, as well as join in discussions, but not all Wikipedians end up equal. There are multiple organizational layers, including "stewards," "bureaucrats," and "administrators." Positions are filled through community-based promotions.

Identity, Affinity, Fun and Games

Contributors to Wikipedia need not toil as nameless members of a virtual Grub Street. Once participants register with Wikipedia (using either their own names or pseudonyms, as is common in blogs), Wikipedians have the opportunity to post short biographies on their "user" pages, complete with photographs and other information they feel pertinent. They may join affinity groups (roughly analogous to the Groups function on Facebook) or receive greetings from the "Birthday Committee" (again, compare Facebook, which offers birthday reminders, typically resulting in a flood of "Happy Birthday" greetings on people's Walls). Reminiscent of the WELL of old, you'll find local Meetup groups, opportunities to Adopt-a-User, and a

"Harmonious Editing Club." In addition to social networking opportunities, there's a Wikipedia "Department of Fun." Categories of "Wikitainment" include contests, games, songs, trivia, and humor.

Acknowledgments

Another important below-the-surface aspect of Wikipedia that makes the project more than simply altruistic community service is acknowledgment for the work one does. Recognition comes in several ways. The simplest is the personal satisfaction contributors gain by seeing their articles (or edits) immediately appear on Wikipedia. When I asked a colleague of mine who is a dedicated Wikipedian why he thought so many people were committed writers and editors, his swift response was, "instant gratification." A more overt form of acknowledgment is having one's entry selected as the "Featured Article" of the day on the homepage of Wikipedia. There is also a formal awards system.[86]

Audience, Accuracy, and Style

Thanks to its legion of contributors, Wikipedia continues to expand. As it has moved to the top of many computer searches, and as an increasing number of academics write for Wikipedia, the project has gained credibility in many quarters. How does it stack up against the three criteria we identified for measuring encyclopedias: content, audience, and style?

The easiest category to address is audience. If you have an Internet connection, you have full access to Wikipedia. In this regard, Wikipedia outstrips proprietary works—either print or online—by a country mile.

The question of content is more complex. Take the issue of completeness. Wikipedia is a work in progress—and projects such as the *Encyclopédie* and the *Britannica* were many years in the making. Since Wikipedians may write on any topic they choose, it's not surprising that coverage is eclectic, with more representation of popular culture and of obscure interests than the breadth of areas (or emphases within each) that a professional editor might have selected. However, it's premature to judge Wikipedia on this score.

The issue of accuracy is a different story. The question here is not just what outsiders think but also how Wikipedia presents itself to potential contributors. Consider these suggestions, from Wikipedia's entry "Contributing to Wikipedia":

Visit Wikipedia: Pages needing attention to find a list of articles by topic.

So far, so good. But now:

> These often need contributions from people who know something about the subject!

What a novel idea: an encyclopedia having authors who "know something about the subject." But the next paragraph is more baffling still:

> Make a list of everything you know. Strike through the things that are not verifiable or not supposed to be covered by Wikipedia. Then, find the proper places to write about the items remaining on the list.[87]

Knowledge as grocery inventory? Wales's heart may be in the right place, but his model of epistemology discards at least two millennia of thinking about the nature of knowledge. Imagine Diderot's reaction.

Such concerns are not meant to detract from the magnitude—and importance—of Wales's endeavors. Undoubtedly, Wikipedia has succeeded far beyond popular expectations in creating a collaborative *tour de force* that enables almost anyone, almost anywhere to get a first look at a subject, where most of the information is accurate, most of the time. Nobody is perfect. Newspapers are continually printing corrections to stories. Publishers used to issue errata sheets even in new works. Book reviewers highlight the errors of authors whose manuscripts have passed through traditional vetting processes. The question about Wikipedia isn't simply whether its articles are accurate but, in Larry Sanger's words, whether we can trust them to be so.[88]

From the outset, critics of Wikipedia feared that composition-by-committee, especially when no authorial credentials are required, undermined the credibility of the project. Sanger left Wikipedia in part because he disagreed with Wales's policy of anti-elitism—that is, giving no special credence to those who had expertise in a field.[89] Robert McHenry described Wikipedia as a "faith-based encyclopedia," where "faith" figures in the assumption that in the give-and-take exchange involved in creating Wikipedia entries,

> some unspecified quasi-Darwinian process will assure that those writings and editings by contributors of greatest expertise will survive; articles will eventually reach a steady state that corresponds to the highest degree of accuracy.[90]

McHenry immediately went on to ask: "Does someone actually believe this? Evidently so."

In late 2005, the British journal *Nature* put the accuracy question to the test, comparing forty-two entries in *Britannica* and Wikipedia dealing with scientific topics such as "Agent Orange," "Ethanol," "Prion," and "Hans Bethe." The average number of inaccuracies per article was very close: around three in *Britannica* and around four in Wikipedia.[91] Not surprisingly, *Britannica* replied to the study with a piece called "Fatally Flawed."[92] The whole discussion brings to mind Einbinder's *Britannica* exposé back in the 1960s. Truth be told, even "authoritative" publications typically have some errors.

Because of the presumed link between issues of correctness and Wikipedia's not vetting authors or editors by scholarly credentials, several alternative online open-source compendia have been launched. One is Scholarpedia, a peer-reviewed encyclopedia edited by Eugene Izhikevich, a neuroscientist in San Diego. Another is Sanger's new project, Citizendium, which he describes as "led" by experts.[93]

As the dust settled somewhat on the accuracy issue, discussion shifted to writing style. Even hard work and good intentions are no guarantee of stylistic success—just ask any writing teacher. Most encyclopedias don't profess to be models of literary excellence. (Recall Van Doren's image of pickled frogs.) In that respect, the *Encyclopédie* and the eleventh edition of the *Britannica* are exceptions. How does Wikipedia measure up?

The notion of style entails many dimensions: level of formality, clarity, and perhaps even grammaticality, not to mention such intangible qualities as elegance. Critics of Wikipedia have taken the project to task for abundant cases of misspelled words, ungrammatical constructions, and tortuous or illogical sentences. Wikipedia's response is that its contributors actively edit entries, so that these sorts of errors get corrected over time.

Moving beyond sentence mechanics, consider level of formality. A recent study weighed stylistic formality in Wikipedia against the *Columbia Encyclopedia*. Criteria included such measures as use of personal pronouns or contractions, average word length, and number of noun suffixes like *-ment* or *-ism*. The analysis found both reference works to be comparably formal.[94]

But now for the more subtle dimensions of style, along with the equally important issue of what should be included. In his article "Can History be Open Source? Wikipedia and the Future of the Past," Roy Rosenzweig aptly observed that "Overall, writing is the Achilles' heel of *Wikipedia*. Committees rarely write well."[95]

Rosenzweig asked how Wikipedia compares with traditional reference sources dealing with history. His conclusions give us pause:

> historical expertise does not reside primarily in the possession of some
> set of obscure facts. It relies more often on a deep acquaintance with

a wide variety of already published narratives and an ability to synthe-
size those narratives (and facts) coherently.... Professional historians
might find an account accurate and fair but trivial.... From the perspective
of professional historians, the problem of Wikipedian history is not that
it disregards the facts but that it elevates them above everything else.

Another way of describing the problem is that Wikipedia precludes assuming
a point of view. Amassing a set of facts is like having the pieces of a jigsaw
puzzle. Possession of the pieces is just the first step. What matters is how you
arrange them.

• • • TAKING SELF-EXPRESSION ON THE ROAD

On the face of things, blogs, YouTube, and Wikipedia look like very different
animals. Blogs are personal written musings, often essentially posted for an
audience of one. YouTube hosts short amateur videos that provide free entertainment for friends and strangers alike. Wikipedia is ostensibly an online
encyclopedia, striving to represent objective knowledge.

Yet all three platforms share the critical feature of affording individuals the opportunity to have their say, potentially before a vast audience. This
opportunity builds on historical precedents (talk radio, home movies, professionally prepared encyclopedias), but allows for a level of freedom of
expression—and participation—that was unheard of even a decade ago.

Wikipedia, YouTube, and blogs have something else in common. All three
have been predicated upon availability of a computer with an Internet connection. For that matter, the same has been true (at least until recently) of
instant messaging and of social networking sites. What happens to language
and social interaction when they go mobile?

7

• • • Going Mobile

Cell Phones in Context

I had just boarded the express train from Heathrow to London's Paddington Station. A sign indicated this was a "Quiet Car," which I interpreted to mean no loud talking, crying babies, or audible music players. Settling into my seat, I observed a large notice advertising a service that would transport luggage from the airport to your London hotel, and then send you a message when your belongings had arrived. The offer sounded good, but how, I pondered, could you receive the message if you hadn't yet reached your lodgings?

It was the summer of 1998, and I, an American, was technologically out of step. As I soon learned, "Quiet Car" meant "no talking on your mobile phone," and those messages about luggage were conveniently (and silently) transmitted over the mobile as well.

For over a century, America had been in the forefront of telecommunications technology. Samuel F. B. Morse's telegraph debuted in 1844. Alexander Graham Bell's telephone was patented in 1876. Universal telephone access was a reality in the United States long before Europeans could make the same claim. Such easy access also came decades before my in-laws in Calcutta finally ended their five-year wait—in 1973—for a phone to be installed in their apartment. Only physicians and businesses were moved faster in the queue, and everyone's line was washed out during monsoon season.

But mobile phone technology left America in the dust. In 1998, I was the proud owner of a car phone, and several colleagues sported rather clunky cell phones (as they are known in North America). Yet by 1998, Europe and Asia were awash in efficient mobile handsets.

First, the European side of the story.[1] Mobile phones—initially, car-based systems—got their start in Sweden in the mid-1950s. In 1969, a Nordic Mobile Telephone Group was established. Soon thereafter, Germany, France, Italy, and Britain independently tried their hand at mobile systems, though costs were high and interoperability across companies was nonexistent.

Enter Groupe Spécial Mobile (GSM), a European consortium formed in late 1982 to create a single mobile telephone system that would function across Europe. It took a decade, but in 1992 eight European countries (Germany, Denmark, Finland, France, the UK, Sweden, Portugal, and Italy) began using the GSM network. Most of Europe had signed on by 1995. The network was designed for transmitting voice calls. Almost as an afterthought, a small amount of leftover bandwidth was made available for creating text messages on the small phone keypad. This feature, known as Short Message Service (or SMS), was initially offered for free in 1993. Over time, GSM began charging for text messages but at a lower price than for voice calls.

Adopted in many parts of the world beyond Europe (in more than 200 countries and territories), by early 2007 GSM accounted for over 80 percent of the global mobile market.[2] To put these numbers into perspective: As of early January 2007, there were 2.73 billion mobile phone subscriptions.[3] More than one-third of the earth's population had mobile access, largely on the GSM system.

Half a world away from the home of GSM, the first car-based Japanese cellular phones were introduced in 1979, with handheld models appearing in 1987.[4] However, not until 1993 did mobile phones experience serious growth, in part because the hefty 100,000 yen subscriber deposit (just over $800 US) was eliminated.

Much as the United States had its historic breakup in 1984 of the telecommunications giant AT&T, Nippon Telegraph and Telephone (NTT) had been a monopoly until 1985. As new rival companies entered the Japanese mobile phone business, their services were based on incompatible standards, reminiscent of Europe before GSM. But one of those services became a standout: NTT DoCoMo ("Do Communications on the Mobile"), which spun off from NTT in 1992.[5] In February 1999, NTT DoCoMo pulled well ahead of the pack with its launch of i-mode, a service connecting mobile phones to the Internet. While Americans were busy sending email and instant messages from their computers, the Japanese were tapping out "short mail" from their mobile phones.

As elsewhere, mobile phones in the United States began essentially as wireless radios, generally mounted in cars.[6] In 1947, two engineers at Bell Laboratories began working on a cellular system for dividing up the radio spectrum to enable many end-users to place calls simultaneously. Three decades later, this "cell" system became the basis for American "cellular" phones.

The first handheld cellular phone appeared in 1984 but, given its size, was more luggable than portable. Over time, handsets shrank, but usership was slow-growing, partly because phones were expensive and so were calling rates. Equally important was the fact that the mobile phone had not developed into a "personal" item in the United States. Given the profusion of landlines, who needed to carry around (and pay for) another phone? What's more, since computers were readily available, if you wanted to send a written message, there was email and later IM.

As American mobile phones finally emerged as a medium-for-the-masses around the turn of the century, they followed a path quite distinct from the GSM world. With GSM, you can use a subscription based in Norway to call the Philippines from Spain. The moment was ripe for the United States to adopt a single transmission platform (and one with global reach) as it made the switch from analog to digital service. Yet the free market prevailed, and the dominant U.S. standards became TDMA—"Time Division Multiple Access" and CMDA—"Code Division Multiple Access." Such service is fine domestically but largely useless once you leave America's shores.[7] Gradually U.S. telecommunications carriers are now beginning to sell GSM phones, but the progress has been slow.

• • • RESEARCHERS TAKE ON THE MOBILE

With more than one-third of the world's population using mobile phones, it's hardly surprising that researchers of all sorts—linguists, sociologists, anthropologists, design specialists, computer scientists, media and communication scholars, policy-makers—have been studying the ways in which young and old, rich and poor have been using these devices. It was the launch of GSM, followed by developments in the Far East (and subsequently in the Middle East, India, and Africa), that put the mobile phone on many scholars' agendas.

A first set of questions explores what people do with their mobile phones: Do they talk or text message? Is the phone for emergency purposes or filling dead time while waiting for a bus? How do patterns differ across age or gender? Also in this category is the linguistic character of communiqués—especially of text messages, which are more easily documented than voice conversations.

A second category deals with social dynamics: How are young people emancipated from direct adult supervision by having access to personal communication devices? Does your mobile phone make a social statement about you through its faceplate, decorative strap, ring tones, or where you place it when you sit down to lunch with friends? Under what circumstances is it appropriate to answer a ringing phone in public or place a call while riding on a commuter train?

"I have to go. I'm getting a better call."

Not surprisingly, the answers to some of these questions vary with cultural—and economic—circumstances. Behavior deemed polite in Spain might be considered the height of rudeness in Japan. Economic necessity in Rwanda generates offbeat mobile phone practices that might leave richer Belgians simply perplexed. To understand mobile phone use, it's often critical to examine the cultural milieu in which the instruments are embedded.

Even a summary of what researchers have learned about mobile phones would take at least a full book the size of this one, so we'll take a different tack. At the beginning of the notes section for this chapter (at the end of the book), you'll find a short list of suggested readings on mobile phones. In the chapter itself, we focus on three themes: a brief sampler of cross-cultural differences in mobile phone practices,[8] a cameo description of usage in Japan,[9] and two pilot studies that Rich Ling and I have done in the United States.

• • • CROSS-CULTURAL SAMPLER

The English drive on the left side of the road, Germans on the right. There's nothing inherent in Vauxhall or VW motivating this difference—just custom.

The Chinese eat with chopsticks; Indians (traditionally), with their hands. Custom again.

Mobile phones are like cars and rice. The practices through which we encounter these items are only partially determined by the objects themselves, with the rest of their functioning often shaped by the culture norms—or pragmatic necessities—of the society in which they are embedded.

Such influences are reflected in mobile phones used in different parts of the world. The Japanese sprinkle their written messages with pictograms (a kitty, a wolf, a sad waif)—sort of emoticons on steroids, which mirror *anime* culture, along with the personalized sticker craze that seized Japan in the mid-1990s, when photo booths for making stickers began cropping up in arcades and shopping centers. By 1997, over 60 percent of middle and high school students (including almost 90 percent of high school girls) collected and exchanged stickers with each other.[10]

Or consider a handy feature available on mobile phones in heavily Muslim countries such as Malaysia. Maxis (Malaysia's primary mobile phone provider) offers handsets that point the way to Mecca. As anthropologist Genevieve Bell explains, for an added fee, "you can also receive regular reminders of *salat* or prayer time customized to your location."[11]

Even in regions that might strike outsiders as rather homogeneous, marked differences crop up in mobile phone behaviors. Consider Scandinavia. When it comes to Sweden, Denmark, and Norway, the languages are sufficiently similar that a Swede can largely figure out what a Dane or Norwegian is saying, or vice versa. Add in Finland, and you have a cluster of cultures known for self-reliance, politically liberal governments, Protestantism, and a traditional fondness for lutefisk at Christmastime. But what about use of mobile phones?

In 2004, the Swedish telecommunications company TeliaSonera undertook a comparison of mobile phone practices among sixteen- to sixty-four-year-olds in Sweden, Denmark, Norway, and Finland. Among their findings: Finns talk nearly twice as much as Swedes, while Norwegians send more than four times as many text messages as their Swedish counterparts. Only two out of ten Danes think it's OK to keep their mobile phones on during a party, while four out of five Swedes are comfortable doing so. Yet only a quarter of Danes hesitate to give out their phone number, while more than 50 percent of Swedes and Finns show caution.

Culture-specific mobile phone usage sometimes springs from economic necessity. Jonathan Donner documents an ingenious practice popular in sub-Saharan Africa of using mobile phones to convey messages without incurring the cost of a call.[12] The practice is called beeping. Say you're a farmer who owns a couple of milk cows. When you have enough milk to sell, you need to arrange to have it picked up by the middleman, who will bring it to market. You (or the members of your family, clan, or village) have a mobile phone, but actually calling the distributor is expensive. And so you beep him. Through a prearranged code, you place the call, ring once, and hang up. The distributor thus knows you have milk available.

The principle of beeping is hardly new. When AT&T still monopolized the American phone network and long-distance tolls remained costly, we all knew how to game the system. If you were traveling and wanted to let the folks back home know you had arrived safely, you fed your coins into a pay phone, dialed the number, rang once, and hung up. You got all your money back, and Mom breathed a sigh of relief. Today, now that domestic long-distance calls are "free" on mobile phones (that is, part of your monthly allotment of minutes), I have yet to hear of Americans using their mobiles for beeping as a thrift measure. However, young people hailing from locations as diverse as Japan and sub-Saharan Africa seem to be re-purposing the beep for a new social function: ring once and hang up to signal (for free) "I'm thinking of you."[13]

Another culturally driven example of mobile usage concerns when to talk and when to send a text message. Carole Anne Rivière and Christian Licoppe collected some interesting data in 2001 and 2002 regarding French customs.[14] Mobile phones became quite popular in France at the end of the 1990s, with GSM's Short Message Service starting in the summer of 2000. Although texting in France was less expensive than voice calls (as in most of the world), the French reserved texting for communicating with a handful of people—generally "intimate" correspondents such as significant others or a very small circle of friends. Most of their mobile phone communications were voice calls—a pattern that looks more American and less European or, for that matter, Japanese.

When the French did send text messages rather than placing voice calls, the decision was often based on social norms regarding privacy while in public space. It's considered rude in France to broadcast private business in public. Text messaging "provides opportunities to communicate with intimate correspondents from public spaces while keeping a proper distance and sense of privacy with respect to bystanders."[15] Anecdotally, I heard a very similar explanation from a Belgian graduate student who was studying in Italy. She found Italians who spoke in public on their *telefonini* to be rude. As a Belgian, she explained, it was far more polite to conduct personal communication via texting when other people were around.

Cellcerts are an example of a culturally generated American phenomenon. I learned about cellcerts from my student Erin Watkins, a devoted fan of the singer Clay Aiken, who parlayed a second-place finish on *American Idol* into a successful recording career.[16] Fans have created a number of online discussion boards devoted to Aiken. What distinguishes some of those boards from your run-of-the-mill fare is that Aiken fans have devised an ingenious concatenation of technologies to bring real-time events to those not physically in attendance.

Let's say that Sarah is at a Clay Aiken concert. By prearrangement, as the concert begins, she makes a cell-phone call to her friend Miriam. Aiken walks out on stage, and Sarah begins narrating what he is wearing, what he's saying, how the crowd is reacting, and so on. Meanwhile, Miriam busily types a version of this narration onto an online Aiken fan site, so a virtual audience around the world can experience the next best thing to being there. The result: a cellcert.

For our final view of mobile phones in cross-cultural context, we turn to Japan.

• • • SILENCE ON THE *CHIKATETSU*

On a lovely Saturday afternoon in 2005, my family was visiting Nara, Japan's first capital and now home to a magnificent park with resplendent temples and shrines, and over a thousand tame deer. The park was filled with families on weekend outings, along with scores of school groups, running the age gamut. At one point, my family had gotten about 100 feet ahead of me, since I had stopped to admire a temple. American style, I called out for them to wait.

Mine was the sole loud voice in the park. Despite the thousands of children around me, only I was shouting.

Fast forward a week to Tokyo Station, through which nearly a million people pass each day. Signs in both Japanese and English admonish travelers, "Don't rush. Keep moving," a near oxymoron that seems to work. I was there to catch a subway (*chikatetsu*) during the morning rush hour, something all my guide books had cautioned me never to do. The Japanese morning and evening commuter traffic raises the saying "packed tightly as sardines" to a whole new plane.

The crowds kept moving—but in silence. Once I squeezed my way onto the Sardine Express, I was again struck by the quiet. Practically no talking (even between people who appeared to know one another), and no mobile phone conversations.

In Japan, the mobile phone is known as *keitai*, literally, 'something you carry with you.' (The same word is used in both the singular and plural, as with all Japanese nouns.) As Mizuko Ito explains, the meaning of the term is firmly embedded within the Japanese cultural context:

> A *keitai* is not so much about a new technical capability or freedom of
> motion but about a snug and intimate technosocial tethering, a per-
> sonal device supporting communications that are a constant, lightweight,
> and mundane presence in everyday life.[17]

In her work, Ito cautions against looking at Japanese *keitai* as a harbinger of
how mobile phones may evolve in the rest of the world. Rather, *keitai* culture
results from "a complex alchemy of technological, social, cultural, economic
and historical factors that make wholesale transplantation difficult."[18]

First, a quick snapshot of how *keitai* are used in Japan. As in much of
Europe, they are commonly employed for sending written messages rather
than speech. Among teenagers and young adults, nearly everyone has a *keitai*,
with ownership tapering off only among older citizens. *Keitai* almost always
carry some sort of physical personalization, which can become quite elaborate among teenage girls. And *keitai* are highly multifunctional: They can be
used for banking, making purchases at grocery stores, doing word processing,
and accessing all manner of Internet functions, including the wildly popular
social networking site Mixi.

Keitai practices are built upon an earlier culture of personal pagers, which
began in the 1980s.[19] In 1987 pagers became more user-friendly when NTT
introduced the first device displaying call-back numbers. Initially, pagers had
largely been restricted to businesses. In the early 1990s, however, subscription prices dropped, and a pager culture developed among young people,
especially high school and college girls. For a while, pager messages only
served as the equivalent of the "I am thinking of you" beep on a mobile
phone. That situation changed in 1995 with the release of the first pager that
could receive text messages. Subscriptions and usage soared. By 1996, almost
49 percent of female high school students in Tokyo had a pager.

In the meanwhile, *keitai* continued to be sold, primarily to businessmen,
with the instrument functioning only for voice calls. Already by the late
1980s and early 1990s, the public had begun to complain about use of *keitai*
in public space. As I had seen in Nara and at Tokyo Station, there are many
circumstances in which Japanese do not speak loudly in public. In Misa
Matsuda's words, "The physical noise was not the problem. Rather, *keitai*
conversations disrupt the order of urban space."[20]

The Japanese solution came in two forms. The first was technological. As
Matsuda explains, in April 1996, DDI Cellular Group (now named "au")
introduced text messaging, making it possible for subscribers to message one
another on their mobile phones. Other carriers followed suit, though initially, messages could not be sent across different carriers (and none was
compatible with the growing GSM standard). In November 1997, J-Phone

(now SoftBank) introduced Internet-based email service, though it was the appearance of NTT DoCoMo's i-mode in 1999 that made written messaging over the mobile phone immensely popular.[21] One explanation for the popularity of messaging on *keitai* is that the new technology offered a socially appropriate way to fill the desire for interpersonal bonding earlier met by pagers.

The second solution was social control. The government began cautioning against use of *keitai* (primarily for talking but sometimes even for written messaging) in certain public places, especially trains (including subways) and buses. Central Japan Railway started announcing on their *shinkansen* (bullet trains) that passengers should refrain from using *keitai* while in their seats. Gradually, announcements were made on other types of public transportation as well. By the late 1990s, scores of newspaper articles were appearing that condemned public use of *keitai*. As of September 2003, passengers began hearing the following kind of announcement on trains and buses throughout Japan, after each stop:

> We make this request to our passengers. Please turn off your *keitai* in the vicinity of priority seating [for the elderly and disabled]. In other parts of the train [or bus], please keep your *keitai* on silent mode and refrain from voice calls. Thank you for your cooperation.[22]

Today, riders find graphic messages posted in train cars, either prohibiting use of *keitai* altogether or at least banning phone conversations. While these announcements do not have the force of law, social custom dictates general obedience.

Regulation of *keitai* use is not an isolated issue. Rather, public manners—here, in Japanese trains—"are part of a broad palette of behaviors that are policed explicitly and persistently by public transportation institutions." As Daisuke Okabe and Mizuko Ito explain,

> Posters illustrate and warn against such transgressions as leaning a wet umbrella on another passenger's leg, eating [food] or applying makeup, groping female passengers, getting fingers pinched by train doors, taking up too much space on seating, or leaving a backpack on rather than holding it at a more unobtrusive level.[23]

The result? Trains in Japanese cities are known for their "precise technical and social regulation and very low rates of disorder, whether it be poor manners, a late train, graffiti, or litter." The usage patterns for *keitai* that have evolved in Japan are all part of a social package.

Do late trains or litter exist? Occasionally. Do people ever talk on *keitai* while riding trains or buses? Indeed they do, although speakers are generally aware they are committing social transgressions. In their observations during 2002 and 2003, Okabe and Ito saw speakers cover their mouths when talking on a crowded train or terminate a conversation if a fellow passenger looked askance. The researchers also heard from young people they interviewed about feeling bad when they took a call, along with stories about users who simply ignored the chastening looks of other riders.

Social norms are reflected in usage statistics. Analyzing 2001 data from research conducted by the Mobile Communication Research Group, Tomoyuki Okada found that youth ranging from middle school to their early twenties sent roughly ten text messages a day. Of the sample, 23 percent said they engaged in "one or two voice calls a day" while 27 percent said "two to six voice calls a week."[24]

Comparing the locations in which people choose voice versus written communication, the usage patterns become more meaningful. Here are the (rounded) percentages of eighteen- to twenty-four-year-olds who used *keitai* for writing or speaking in the following locations:[25]

	Written Message	Voice Communication
Home:	93%	83%
School:	39%	32%
Restaurant/Café:	21%	13%
Street:	48%	49%
Station/Bus Stop:	31%	24%
Train/Bus:	36%	5%

It's OK (according to some people) to talk while waiting for a train or bus, but not OK once you get on board.

The Japanese are nuanced users of *keitai* for controlling the ways in which they interact with others. One housewife deftly adjusted her own "volume" by deciding when to call and when to send a text message:

There are things in this world that can only be said in e-mail [= text message]. Like when I get into a fight with the children, some things are difficult to say on the phone. So I write, "I'm sorry, I didn't mean what I said." Then they write back and say, "I'm sorry, too. I also went too far." On the phone, they would say, "Stop nagging."[26]

Volume control also occurs when *keitai* users screen incoming voice calls by looking at caller ID before deciding whether to answer. Results from the 2001 data indicate that almost half of unmarried users screened their incoming calls "very often" or "sometimes."[27] As Matsuda explains, this proportion is not surprising, given the large number of people to whom young users commonly give out their *keitai* numbers.

Finally, *keitai* bestows a sense of control over when you can contact other people. In the Mobile Communication Research Group study, 25 percent of the respondents said that in having a *keitai*, they "often" felt that "there were fewer occasions where I get irritated because I can't get through to somebody."[28] However, *keitai* can be a double-edged sword: They give control not just to you but to potential interlocutors as well. As a male Japanese college student observed, "It's not that I want to be connected, it's that I want to be able to make a connection."[29] As we'll see, American mobile phone users appear to agree wholeheartedly.

• • • MOBILE PHONES AMERICAN STYLE

Just as Muslim, Scandinavian, sub-Saharan African, French, and Japanese mobile-phone practices grow out of specific cultural contexts, cell-phone culture in the United States is the child of both its technological and social roots. Before Americans became enamored with cell phones, they were already a nation of talkers and typists. These experiences have colored their choices about what to do with the new technology.

A Nation of Talkers

Speaking is a defining point of being human. Do Americans talk more than others? Hardly, but it's fair to say that a lot of them feel no qualms about talking in public—even loudly. That comfort level in ignoring the volume control extended to use of the telephone.

Much as Americans got an early start in their love affair with the automobile, they embraced the telephone from its inception. Lacking a phone at home (or if you were traveling), you could always find a pay phone at the local drug store. As the country became wealthier after World War II, extension handsets and multiple phone lines sprang up in homes across America. Expansion of unlimited calling service for a monthly fee, along with

"I have to hang up now. You just walked through the door."

gradual reduction in long-distance rates, meant that average callers (not just teenage girls) might be on the phone for hours. The United States was light-years apart from the pronouncement made in 1895 by the British Post-master-General that "Gas and water were necessities for every inhabitant of [England]. Telephones were not and never would be."[30]

When small, reasonably priced cell phones began making their way into the U.S. marketplace by the late 1990s, Americans knew just what to do with them: talk. Given the American pricing system, the decision was highly ra-tional. Until very recently, all mobile phones in the United States worked on a monthly subscription basis, generally tied to a two-year contract. You paid for a maximum number of minutes of talk time, so you might as well use them. When text messaging made its gradual debut, few people were interested. It seemed unnecessarily cumbersome. What's more, you needed to pay extra for each text message. Computers—readily available at home, in the work place, in libraries—were the natural choice for electronically-mediated communi-cation. Email and instant messages were "free"—once Internet access was paid for. No wonder Americans kept on talking, not texting, on their mobiles. Writing was something you did with ten fingers on a keyboard.

A Nation of Typists

When typewriters became commercially viable in the 1880s, a new profes-sion sprang up. Armies of women were trained to operate the machines, which soon became office fixtures. Over the decades, others began typing as well—journalists, authors, college students—but frequently relying on the hunt-and-peck method. I still remember the signs posted around my undergraduate campus by those who actually knew how to touch type (usually females), offering their services at twenty-five cents a page to those (typically males) for whom laboriously typing a ten-page paper would take the better part of a day.

Steve Jobs and the Apple IIe helped change all that. With the introduction of personal computers into the classroom in the 1980s, school children began finding their way around computer keyboards—first hunt-and-peck (like most of their parents), but then fortified by classes in "keyboarding skills." As these children progressed through the grades, they became as comfortable at a computer keyboard as they were with a television remote control. Once in college, they owned a personal computer or had ready access to computer labs. Everyone typed away on school assignments, email, and instant messages.

By the early years of the new century, America was firmly a computer culture, especially in comparison to most other countries. As of 2005, there were 76 personal computers per 100 inhabitants in the United States, while the comparable figure for Europe was 31.[31]

By 2006, 73 percent of American adults were on the Internet.[32] Email remains the most popular Internet application, with a usership of over 90 percent of online adults.[33] Among teenagers and young adults, IM is particularly widespread. As of 2005, 42 percent of adults who went online used instant messaging, along with 75 percent of online teens.[34]

At the same time, Americans have not abandoned their comfort with the telephone. Given a choice of technologies for communicating with friends, 24 percent of American teenagers chose IM, while 51 percent preferred landline phones, 12 percent opted for voice calls on mobile phones, 5 percent selected email, and only 3 percent used text messaging.[35]

When we compare texting across countries, the United States remains on the low end of the spectrum. As of 2006, approximately 70 percent of Norwegians aged nineteen to twenty-four reported daily use of text messaging.[36] By contrast, in the United States, as of 2005, approximately 4 percent of all people and 18 percent of those aged eighteen to twenty-four used texting on a given day.[37] A study released in December 2006 of young people aged thirteen to twenty-four highlights the discrepancy between Europeans and Americans. Among mobile phone subscribers from Germany, Italy, Spain, and the UK, between 81 and 86 percent sent text messages. In the United States, the number was 39 percent.[38]

I vividly remember a heated discussion among students in one of my classes in late 2002 over what a mobile phone was good for. About a third of the students owned mobile phones at that point. However, only one of them—who had just returned from a semester studying in Italy—knew about text messaging. As she explained how convenient it was to tap out messages with your thumb on the small keypad, the other students rolled their eyes in disbelief: Why would anyone struggle to produce a text message when you could simply call (or use a computer to send an email or an IM)? Such was the mind-set of average American college students at the time mobile phones— and eventually text messaging—began to be aggressively marketed in the United States.

Beyond convenience, never underestimate the power of money. In Europe, text messaging has generally been less expensive than voice calls, accounting in large part for the European explosion of texting among the young. In the United States, text messaging is an added expense (on top of voice contracts), paid either by the message or through a monthly messaging plan. The choice between a voice call or a text message is sometimes determined by who is paying the bill. Parents of teenagers or college students know to expect monthly voice charges but sometimes balk at additional fees for texting.

American Mobile Phone Practices: A Pilot Study

In collaboration with Rich Ling, my students and I investigated mobile phone practices on two college campuses in the United States in late 2005. Ours was essentially a pilot study, without pretensions at representing all university students, much less all Americans. (Since we knew of no prior studies, we had to start somewhere.) Our actual questionnaire included many items, but here we'll focus on four issues: (1) personalization of mobile phones, (2) text messaging versus voice calls, (3) reasons for using a mobile phone, and (4) attitudes toward the technology.[39]

Personalization of Mobile Phones

Phones can be personalized through physical decorations or distinctive ring tones. In Japan, for instance, mobile phone users of both genders attach decorative straps to their handsets. These accoutrements are widely sold at tourist sites and convenience stores. Phones can also be personalized with pictures or stickers (again popular in Japan), or a fashionable faceplate. Casual observation of mobile phone users in the United States suggested little use of physical personalization. A 2005 study of college students in suburban Philadelphia asked a related question: How important is it to you that your mobile phone be "up-to-date"? The answer "not very important" was given by 43 percent of the women and 63 percent of the men.[40]

Downloaded ring tones are an audible form of personalization. Estimates for global ring tone sales in 2005 were over $4 billion, with American sales for 2006 projected at about $600 million.[41] Users may choose an individual song (or series of tones) for all incoming communications or distinctive rings for different people in their address book. Given the relative novelty of mobile phones (and of pay-per-download ring tones) in the United States, we were curious how prevalent ring tone personalization had become.

Text Messaging versus Voice Calls

Of those who used text messaging, we wanted to know how long they had been texting. Did they have a monthly messaging plan? How many texts did they send or receive daily? Among texters, how many of their mobile phone communications were voice calls and how many were text messages? In both cases, what was the physical proximity and the identity of their interlocutors?

Reasons for Using a Mobile Phone

Why, we wondered, did students decide to talk rather than text. Beyond texting and talking, mobile phones may serve other functions, such as cameras, calculators, alarm clocks, or platforms for music or games. A further use suggested by students helping design the study was "pretending to talk when actually you are not."

Attitudes toward Mobile Phone Use

Finally, we were interested in user attitudes regarding mobile phones. In what physical locations did college students commonly make voice calls, and were the students bothered when other people spoke on mobiles in public? Additional queries focused on literal volume control: Did survey participants feel they spoke more loudly on a mobile phone than when speaking face-to-face? Did they believe other people spoke more loudly? A last question asked what college students liked most and disliked most about their mobile phones.

Survey Respondents

Our sample included 93 college undergraduates: 34 females and 34 males at American University (AU), along with 25 females at the University of

Michigan. The mean age was between 19 and 20 years old. Everyone had a mobile phone. At AU, both males and females had owned phones about three-and-a-half years, while the Michigan females averaged over four years. (The range in both groups was vast: from a couple of months to eight years.)

What We Found

Personalization of Mobile Phones

Only one out of seven decorated their phones. At AU, more than three times as many females used decorations as males did. Michigan females were less into decorations than their female counterparts at AU.

The comparatively higher proportion among AU females (one out of five) might reflect regional fashion or the fact that many students at AU had traveled abroad and encountered decorated phones. Anecdotally, males at AU were perplexed at the idea of adorning a phone. One student recounted his surprise at finding that a male Japanese co-worker had a novelty strap attached to his phone, which the American had judged to be effeminate.

This relative lack of decoration may also reflect American college students' comparative lack of interest in the phone as a form of personal expression. Most phone sales in the United States are tied to extended voice-call contracts, so models are not easily changed on a whim. Many students acquire their phones through family plans (including buy-one-get-one free), where parents have the final say in selection. When students in the United States do procure trendy models, they seem not to adorn them, letting the design of the phone speak for itself.

About half the students downloaded ring tones enabling them to distinguish between callers (say, a significant other versus everybody else). As a group, females—especially at Michigan—were slightly more involved in using distinctive ring tones than males. Among those using distinctive tones, the average number of ring tones downloaded ranged from about four to seven. Males (at AU) were at the high end of the scale, with Michigan females at the low end.

Text Messaging versus Voice Calls

Are American experiences as talkers (on landline phones) and as typists (on full computer keyboards) reflected in the way college students are using mobile phones? We approached this question by comparing the amount of text messaging versus voice calls among participants who did texting. Nine out of ten students in the AU sample had tried—or actively used—text messaging. The number at Michigan was closer to 96 percent. The average person who used texting had been doing so for nearly three years, though there was huge variation across individuals, both in how often they text messaged and how long they had been at it.

Monthly texting plans are clearly taking off in the United States. Almost 60 percent of the AU subjects—and over 90 percent of the Michigan females—subscribed to a monthly texting plan, rather than paying by the message. This discrepancy between students at the two institutions might reflect the fact that Michigan students had owned their phones longer and had more extensive texting experience. Alternatively, texting might have been more fashionable on the Michigan campus—or the sample simply unrepresentative.

Daily inbound traffic of text messages was about the same as outbound. Our students at AU reported sending (and also receiving) three or four messages each day. The Michigan women were somewhat more prolific, averaging between five and six messages sent and also received. Given that the Michigan crowd were more likely to have monthly texting plans, it isn't surprising that they used them.

We asked participants how many of the last ten mobile phone communications they had initiated or received were voice calls and how many were text messages. For every three text messages, there were seven voice calls. (In comparison, as we saw earlier, Japanese teenagers and young adults sent an average of ten written messages a day, but generally less than that number of voice calls per *week*.) If texting volume in the United States grows in the future, the amount of talking won't necessarily diminish. Given the huge number of voice minutes most students have on their phones (sometimes 500–1,000 per month), there's no reason for increased texting to diminish voice-call traffic. This situation is obviously different in most of the world, where the majority of young people use plans deducting charges from the same SIM card for each voice call or text message.

Anecdotal evidence suggests that removal of price barriers causes some young people outside the United States to shift the balance to favor talk. In April 2006, students at the University of Udine in Italy informed me that for the past few months, they had been making voice calls at least as often as sending text messages because a local service provider was running a voice-call promotion. They anticipated reverting to more texting once the promotion ended.

Returning to our study, where, geographically, were the communication partners who were receiving calls or texts? The preponderance of voice calls were made to people within five miles of the caller (but not within the same building or within eyesight). For most students, the next most prevalent group to call was people more than thirty miles away. The same distribution held true for recipients of text messages.

Who were these people on the receiving end of calls and text messages? Of the last ten voice calls our AU students had made, roughly 60 percent were to same-aged recipients. The next most prevalent group was parents—approximately 20 percent. Voice traffic to siblings was about one out of ten. Subjects at Michigan were also most likely to speak with same-aged cohorts and next most likely to call parents.

With texting, same-aged comrades were again the most common recipients of messages. At AU, the next most likely recipients were siblings, with negligible numbers of messages sent to parents or "others." Our AU males were more likely than AU females to text-message people of their same age. At Michigan, the ordering among the three categories (same-age, siblings, and parents) remained the same, but nearly a third of all text messages were sent to "others."

We suspect that a substantial proportion of calls and texts sent over thirty miles away went to siblings, parents, and "others" (such as potential employers). Our subjects were twice as likely to call parents as to call siblings, but (at least for AU subjects) more than five times as likely to send text messages to siblings than to parents. While American parents are steadily increasing their familiarity with text messaging, the numbers nowhere approach those of their progeny.[42]

Reasons for Using a Mobile Phone

We asked students to rank order their top three reasons for making a voice call or composing a text message. While we didn't specify the recipients of these calls or text messages, the majority (as we now know) were same-aged friends in the nearby vicinity.

Reasons for making voice calls were fairly consistent. By far, the most important motivation was "keeping in touch," followed by "arranging to meet in a few minutes" and "arranging to meet in a few hours" (a close third). Slightly farther behind were "sharing news" and then "killing time while waiting or traveling," with "asking advice" distantly bringing up the rear. Results for text messaging followed a different set of priorities. "Arranging to meet in a few minutes" was first by a nose-length, with "arranging to meet in a few hours" and "sharing news" tying as very close runners-up. Not far behind were "killing time while waiting or traveling" and then "keeping in touch" (which, as we saw, was first for voice calls), with "asking advice" again a distant last. To summarize:

Voice Call Rankings

1. keeping in touch

2. arranging to meet in a few minutes

3. arranging to meet in a few hours

4. sharing news

5. killing time while waiting or traveling

6. asking advice

Text Messaging Rankings

1. arranging to meet in a few minutes

2. arranging to meet in a few hours*

3. sharing news*

4. killing time while waiting or traveling

5. keeping in touch

6. asking advice

*tied

What this summary camouflages is marked differences between genders and between the two campus settings. For voice calls, the variation was relatively subtle. Unlike their male classmates, AU females were more likely to call to "share news" than to make arrangements to meet. By contrast, Michigan females called more often to "kill time while waiting or traveling" than either to make arrangements to meet in a few hours or to share news.

Variations in reasons for texting were starker. "Sharing news" was very important for AU females (ranked first), but far down on the list for AU males (ranked fifth), with Michigan females in between (tied for second). Remember that AU females also ranked "sharing news" much more highly in making voice calls than either AU males or Michigan females. AU males and Michigan females both found longer-term meeting arrangements ("in a few hours") to be the most important function of texting, which was far down on the list of AU females (ranked fifth). Everyone agreed, though, that texting wasn't a good vehicle for seeking advice—mirroring their judgment regarding voice calls.

We also asked why students sometimes decided to text message rather than talk on the phone. Their three top reasons, in order, were:

1. It's not a good time for me to talk
2. I want to make my message short, and talking takes too long
3. It's not a good time for the recipient to talk

These answers were highly pragmatic. In first place, the sender's own convenience; in third place, the convenience of the recipient. There's a certain logic here, since initiators of communication always knows their own circumstances but often not the availability of the person they are trying to reach.

Ranked second was "I want to make my message short, and talking takes too long." This was a theme many students reiterated anecdotally. Throughout this book we have suggested that contemporary information communication technologies increasingly enable us to control the terms of conversational engagement. Dispatching an email rather than making a phone call often saves time, because the sender can dispense with most social pleasantries and keep the message on-topic. These college students likewise sought conversational control by choosing texting over talking on mobile phones.

Beyond interpersonal communication, mobile phones serve other functions: as a clock or alarm (nearly all our subjects used it), a calculator (about half used it), and music player or game platform (males were heavier users of both than females). For those with camera phones, we learned it was more common to take pictures than to send them.

But my students had also pleaded to ask another question, which seemed at odds with the raison d'être of mobile phones: "Do you ever hold your phone to your ear to pretend you are talking, when actually you are not? If 'yes', in which situations?" The main options they came up with were "when I'm trying to avoid talking with someone I see" and "when I'm out alone at night (e.g., on the street, in a bus) and feel uncomfortable."

It turned out that a substantial proportion of respondents sometimes pretended to be talking on their mobile phones: slightly more than one-third of the students (male and female) at AU and slightly more than two-thirds of the females at Michigan. The reasons for such pretense varied by gender. Females (both at AU and Michigan) were more than twice as likely as males to act as if they were talking in order to avoid other people. Females were overwhelmingly more likely to feign conversations than males when alone at night. Informally, female students explained they felt less likely to be harassed (or attacked) if their potential predator saw they could immediately summon help.

Attitudes toward Mobile Phone Use

Our last questions tried to gauge students' attitudes toward mobile phone use—their own or other people's. We began by asking where subjects used their mobile phone for voice calls. For nine out of ten of student subjects at both AU and Michigan, talking and walking went hand in hand. By comparison, only 49 percent of the Japanese had reported talking on the street. Around 40 percent of our participants were comfortable making voice calls while on public transportation—in comparison with the tiny 5 percent of Japanese age-mates. Clearly, America lacks Japan's rigorous exercise in socialization.

Restaurants make for another interesting comparison. In the Japanese sample, only 11 percent of females and 16 percent of males reported making voice calls in restaurants or cafés.[43] In our data, there were considerable differences of opinion. At American University, 15 percent of our female participants believed calling in restaurants was fine, compared with 29 percent of males. However, a whopping 42 percent of the women in our University of Michigan sample indicated that calling from eating establishments was OK. The differences between AU and Michigan could reflect skewed samples—or the fact that AU is located within an urban metropolis, while the University of Michigan dominates a small college town, where businesses catering to students have been the lifeline of the economy.

Were participants bothered by other people's voice calls in public? Our data here are just from AU. Despite the willingness of so many students to talk on their phones in public settings, 77 percent of the females and 48 percent of the males responded "yes" to the question "Does it ever bother you when other people talk on their cell phones in public places?" Reasons included excessive volume, overhearing other people's private conversations, and phones being used in inappropriate places (such as in church or in a restroom).

A related issue was whether students spoke more loudly on cell phones than when face-to-face, and whether they felt other people did so. Once more, there was a mismatch between genders: 35 percent of the females from AU and 79 percent of the males (again, at AU) admitted to speaking more loudly when on a mobile phone. However, 70 percent of both genders at AU felt that "other people" spoke more loudly on mobile phones than in face-to-face conversation.

Our last question asked what students liked most and what they liked least about their mobile phones. Fully half the AU subjects—males and females alike—volunteered that what they liked most about their phones was always being able to reach people. At the same time, the most common complaint (especially among AU males) was that other people could always reach you. This Janus-faced sentiment is reminiscent of the male Japanese college student who said that "It's not that I want to be connected, it's that I want to be able to make a connection."

Reflecting on the Findings

One theme weaving through our data is gender-based differences in the way our sample of American college students were using mobile phones. Earlier we talked about divergences in how males and females write instant messages.

The issues with mobile phones seem driven not by implicit standards of correctness (as with IM) but by fashion, safety, and social networking.

Females in our study physically decorated their phones more than males, suggesting that at least some women are using their phones for presentation of self or to make fashion statements.[44] While both males and females feigned conversation to avoid encountering someone they saw, such control over social interaction was far more prevalent among females. (It would be interesting to compare this behavior with other social avoidance mechanisms such as downward gazes, defensive body postures, or intentionally crossing the street or looking the other way.) A third distinction, only present in the AU data, was the importance of making voice calls or sending texts to share news—ranked considerably more highly by AU females than AU males. Analysis of discourse functions on landline phones, on IM, or face-to-face would help contextualize this finding.

Differences in responses between AU and Michigan remind us how important it is to be on the lookout for sample bias. Although only a few distinctions between AU and Michigan reached statistical significance, the females at Michigan shared several characteristics setting them apart from all students at AU and specifically from AU females. Our Michigan subjects had owned their mobile phones longer, had more experience with text messaging, were more likely to have texting plans, sent and received more texts daily, and were far more likely to choose texting over talking for the express purpose of keeping their messages short. The Michigan students were also more comfortable talking on mobile phones in restaurants and somewhat less likely to be bothered by other people talking on phones in public.

Several hypotheses might explain the different institutional patterns. AU, a midsized private university, sits at the edge of Washington, DC, and student phone habits may be influenced by the larger city, in which they spend a lot of time. The campus also has a high proportion of international students, and many of the Americans study abroad—and observe non-U.S. mobile phone practices. Michigan, a large public university in the Midwest, comprises a substantial proportion of the town's population. Customs may be more directly shaped by students, and less by cosmopolitan standards or international experience. The Michigan women might also represent a less balanced sample than those at AU. Since nearly all Michigan participants were enrolled at the time in a course on mobile communication, members of this self-selected group might be more seasoned mobile phone users than their average classmates.

What about the lopsided ratio of voice calls to text messages that we found in our study? Stark differences in pricing plans between the United States and elsewhere make it difficult to separate cultural practices from economic exigencies. These issues may become disentangled only if the cost of voice calls declines outside the United States or if text messaging costs are folded into American voice plans.

In the mid-2000s, mobile phones in the United States are a technology in transition. While both ownership of mobiles and use of texting continue to increase, mobile technology got a late start in America. This fact, along with its distinctive history with universal access to landlines and ubiquitous personal computers, may lead the country's mobile phone usage on a different path than found in Europe and Asia.

Some of these differences result from cultural practices that go beyond the mere presence of technology. For more than a decade, most American school systems have encouraged students to prepare written work on computers rather than writing papers longhand. This directive further reinforces the comfort level American young people feel in using ten fingers on a full-sized keyboard to create electronically-mediated communication, rather than eagerly embracing small mobile phone keypads. Similarly, American culture is generally permissive of engaging in private talk in public places. Talking on the phone in airports, restaurants, and alas, even classes is perceived in the United States as less of a social faux pas than in many other parts of the world.

• • • TEXT MESSAGING AMERICAN STYLE

In 1455, when Johannes Gutenberg produced his Great Bible of Mainz, the book was a dead ringer for a medieval manuscript, complete with hand-drawn graphic illuminations and the "black letter" style of script popular for religious tracts in Germany at the time. When Steve Jobs put trash cans and file folders onto the virtual desktops of Apple computers, his goal was to give the machine the familiar look and feel of an office. When email became a mode of online communication, the template came straight out of memoranda: "To," "Subject," and "cc," with the contents of "From" and "Date" automatically generated.

Creators of new technologies often consciously build in features making the novel contrivances seem familiar. At the same time, users sometimes carry over anachronistic behavior patterns that made sense only before the new invention came along. Think of the facial expressions and hand gestures so many of us employ when talking on the telephone—obviously wasted effort. Or consider the fact that we generally sign our emails (following traditional letter format), even when our names are routinely displayed at the tops of the messages.

Text messaging is a relatively new technology for Americans. Yet these same users are generally veterans of writing messages on computers through chat, listservs, email, and, especially for young adults, IM. As we have seen, American college students follow rather predictable language patterns in doing IM, influenced in part by the fact they use their same ten fingers on the same computer keyboard to prepare formal written school work. Given users' experience with constructing electronically-mediated messages, are there traces of these writing habits (specifically as seen in IM) in the new written genre of text messaging?

Obviously, measuring text messaging against IM is something like comparing apples and oranges. IM is done with ten fingers; texting input takes one or two thumbs. IM conversations are often in the background of other computer-based activities, while texting is more a stand-alone activity. IM is "free" (once you have an Internet connection); text messages have a price attached (either by the message or for a monthly plan). IMs can be as long as you like, in comparison with text messages (at least in the United States), which are limited to 160 characters. Instant messages are often chunked into seriatim transmissions, yielding a sequence of IMs that together constitute an utterance ("that must feel nice" + "to be in love" + "in the spring"). By contrast, text messages are overwhelmingly composed all of a piece and sent as single transmissions.[45]

Another important consideration is the access people have to texting on mobile phones and to IM on computers. Compared with many parts of the world, texting is a new technology in the United States. Conversely, Internet access on personal computers (which can send, among other things, instant messages) has been more readily available in the United States than in most other countries. According to Internet World Stats, 70 percent of Americans have Internet access, compared with 37 percent of Europeans.[46] Use of IM on computers is steadily growing in Europe. In 2005, for example, *Le Monde* reported that about a third of young people between ages twelve and twenty-five were doing IM.[47] Rich Ling notes that in Norway, between 40 and 50 percent of those between the ages of sixteen and twenty-four use IM.[48] My own studies of mobile phone use in Sweden (as of late 2007) suggest that at least 80 percent of college students make regular use of IM. However, it's a fair bet that the United States outstrips the vast majority of the world.[49]

Detailed analyses of texting have appeared for several languages, including German, Swedish, Norwegian, and British English.[50] Among the stylistic features noted are abbreviations, acronyms, emoticons, misspellings, and omission of vowels, subject pronouns, and punctuation. Since texting in the United States is comparatively new, collecting texting data has hitherto been difficult.

By contrast, American IM has been more amenable to research. We've already seen that abbreviations, acronyms, and emoticons are less prevalent in American college student IM conversations than suggested by the popular press. To move beyond media hyperbole regarding text messaging, we need corpus-based analyses of texting. By collecting data from similar populations, we can compare the linguistics of texting and IM.

Why compare these media? If parents or teachers are concerned that IM and now texting are destroying young people's grasp over the written word, we need firm information on the linguistic features of their messages. Is texting ruinous but not IM? Or vice versa? Without data, we can't even approach the issue. But less judgmentally, a comparison may offer cultural insight regarding the relationship between language technologies and writing practices. We know that American college freshmen arrive on campus as proficient typists, accustomed to using their computers for both writing papers and doing IM. The question now is whether American texting is colored by prior experience with IM, which in turn reflects proficiency with word processing.

Another Pilot Study

Since there were no prior linguistic analyses of American text messaging that we were aware of, our first goal was to start mapping the territory. We knew to look at the standard landmarks: message length (in characters and in words), emoticons, abbreviations, and acronyms. But there were other linguistic features that intrigued us as well, stemming partly from my earlier findings with IM and partly from our informal observations of student text messaging. And so we also analyzed the number of one-word messages, how many sentences there were per text transmission, use of contractions versus full forms, and some of the nitty-gritty details of punctuation. Using these same questions, we examined an IM corpus—and compared the results.

Texting and IM Data: Back to Michigan and AU

Following a protocol commonly used in mobile phone studies elsewhere in the world, Rich Ling distributed paper diaries to twenty-three female students at the University of Michigan and asked them to record exactly all the text messages they sent over a twenty-four-hour period. We ended up with a total of 191 distinct messages, made up of 1,473 words.

Our IM conversations were drawn from the study described in chapter 4. Admittedly, it was hardly ideal to compare IMs collected in spring 2003 with texting messages gathered in fall 2005, particularly because they were from different subjects. Since the IM data were at hand, however, they at least offered a first comparative look at IM versus texting among American college students. To create a matched IM sample, I drew a random sample of 191 IM transmissions, using only female data. The resulting sample consisted of 1,146 words.

What We Found

The first part of our analysis looked at length issues (number of words, number of characters, and one-word transmissions). Next we examined emoticons and different sorts of lexical shortenings, including abbreviations, acronyms, and contractions—and use of apostrophes in contractions. The last category dealt with sentence punctuation: overall punctuation, use of question marks and periods, and creative applications for ellipses and dashes.

Here are the overall findings.[51]

Length

Message length was significantly different between texting and IM. Text messages averaged 7.7 words; IM transmissions averaged 6.0. For IM, we need to keep in mind that sequencing of consecutive IMs is very common. While the average text message was longer than the average IM transmission, the overall average length of a complete conversational turn (combining sequences of seriatim IM transmissions) was longer in IM.

Not surprisingly, the average number of individual characters (including letters, numbers, and punctuation marks) per transmission in the texting data was also significantly larger than in IM. Text messages averaged almost 35 characters each; the average number for IMs was just under 29 characters.

Another factor contributing to message length is the number of one-word transmissions. There were significantly fewer one-word text messages (7 out of 191, or nearly 4 percent) than one-word IM messages (36 out of 191, or almost 19 percent). This difference probably reflects the fact that while individual text messages have a cost attached to them, IM transmissions don't.

Words in text messages were slightly shorter than those in IMs: an average of 4.6 characters per word in texting and 5.0 characters in IM. One explanation is linked to abbreviations. There were more word truncations (such as *yr* for *your* or *can't* for *cannot*) in texting than in IM. Another factor is word choice. Texting style is more likely to use everyday words like *meet*

(4 characters) or *happy* (5 characters), while some IMs wax political or philosophical, with attendant vocabulary to match, such as *encounter* (9 characters) or *entitlement* (11 characters).

There were also differences in word variety. The IMs contained a broader vocabulary than the text messages. For IM, there were 480 distinct "types" of words (with a total of 1,146 "tokens"—that is, words in the entire sample), yielding a type/token ratio of 0.42. The texting corpus contained 405 distinct word types (with a total of 1,473 words), giving a type/token ratio of 0.28. The higher the ratio, the more different vocabulary items the students used.

Next we looked at the average number of sentences appearing within a text message or IM. Because commas and periods were sparse and sometimes haphazard, we couldn't always rely upon standard punctuation to help distinguish between stand-alone sentences and elements of larger sentences. Accordingly, we created guidelines for what counted as a sentence. We identified all the words that potentially could stand as full sentences (such as *ok, haha, whatever, seriously*) and the words that looked like components of a preceding or following sentence (for instance, *hey Jane, besides, hon.* [=*honey*]). Our essential criterion was whether, in our impressionist judgment, the candidate commonly appeared as a stand-alone word (or phrase) in the speech of young adults. While this measure may overestimate the number of sentences per message, the coding scheme was applied consistently to both texting and IM.

To illustrate how the number of sentences per message was calculated, here's a sample transmission from the IM corpus:

> alright, hon. I am exhausted, I think I am going to be the biggest dork alive and go to bed

We divided this IM into three sentences:

> sentence 1: alright, hon.
> sentence 2: I am exhausted,
> sentence 3: I think I am going to be the biggest dork alive and go to bed

The texting corpus contained significantly more multisentence messages than the IM corpus. Almost 60 percent of the text messages had more than one sentence, compared with only 34 percent of the IMs. The average number of sentences per text message was 1.8, compared with 1.3 for IM. Given the possibility in IM (but not texting) of sending sequences of messages seriatim without added cost, this finding isn't surprising.

Emoticons and Lexical Shortenings

Emoticons were very infrequent in both texting and IM. In the texting sample, only 2 were emoticons. (Both were smileys at the ends of messages.) For IM, there were only 5 emoticons: 4 smileys and 1 frowny face. In three instances, the emoticon constituted the entire transmission.

Acronyms were equally sparse in both texting and IM. In texting, only 8 acronyms appeared: 5 cases of *lol* ('laughing out loud'), and 1 each of *ttyl* ('talk to you later'), *omg* ('oh my god'), and *wtf* ('what the [expletive]'). In IM, there were just 4 acronyms: 3 examples of *lol* and 1 of *ttyl*.

What about abbreviations? In IM, there were no clear-cut examples of abbreviations specific to online communication. One example of *b/c* for *because* occurred, but this same abbreviation is regularly found in informal writing and predates email or instant messaging. There were also three miscellaneous lexical shortenings: *ya* (*you*), *prob.* (*probably*), and *em* (*them*). However, these lexical forms commonly appear in the informal speech of many American college students, so we can't presume these are abbreviations specifically chosen for use in IM.

By comparison, texting had a substantial number of abbreviations. Among the 47 unambiguous cases, there were 26 instances of *U* (*you*), 9 uses of *R* (*are*), 4 examples of *k* (*OK*), 6 occurrences of *2* (*to*, both as a single word and as part of the word *today*), and 2 instances of *4* (*for*, both as a single word and as part of the word *before*). While this number is far greater than what we found in the IM corpus (namely none), a total of 47 clear abbreviations out of 1,473 words is hardly overwhelming.

Besides these obvious abbreviations in texting, there were several examples involving vowel deletion: 2 instances of *b* (for *be*), and 1 each of *latr* (*later*) and *ovr* (*over*). It's difficult to be certain whether these examples represent intentional lexical shortenings (a phenomenon described in Swedish text messaging),[52] laziness, or simple typing mistakes.

Finally, there were nearly a dozen cases of miscellaneous word shortenings. Some of them appear to be texting shortcuts, such as *Sun* (*Sunday*) and *tomm* (*tomorrow*), while others can also be found in casual speech (*ya* for *you* and *cig* for *cigarette*).

Moving beyond the usual round of suspects, we turned to contractions. Earlier, we made much of the fact that in their IM conversations, females only used contractions 57 percent of the time that such abbreviated forms were linguistically possible. This, despite the fact that contractions are extremely common in American informal speech.

Beyond the contractions themselves is the issue of punctuation—an apostrophe—within the contraction. In both IM and text messaging, the apostrophe appears on the screen as a single mark of punctuation. Yet the steps necessary for creating apostrophes are sharply distinct. In IM, the apostrophe (on standard keyboards) requires only a single stroke of the little finger on the right hand—not even use of the shift key. By contrast, creating an apostrophe in text messaging requires at least four key taps (depending upon your phone). Omission of an apostrophe in IM constitutes a simple form of shortening, while in texting, such exclusion eliminates a considerable amount of work.

In scoring contractions, we looked at use of uncontracted forms (*do not*) versus contracted forms (*don't*), and then calculated percentages against the total potential contractions. For scoring apostrophes, we looked at absence or presence of the mark for all contractions in the respective samples. (We didn't examine possessives such as *Maria's*.)

Starting with use of contracted versus uncontracted forms: There were more contractions in texting than in IM. In texting, 85 percent of all potential contractions were contracted, which was significantly higher than the 68 percent for IM. (We mentioned a moment ago that for the larger IM corpus described in chapter 4, females used even fewer contractions: 57 percent of potential uses. Such are the vagaries of sampling.)

When it came to use of apostrophes in contractions, the differences between texting and IM were even more striking. Far fewer apostrophes were used in texting contractions than in IM contractions. In the texting sample, there were 72 contractions. Apostrophes were only used in 23 of these (32 percent of the time). By contrast, in the IM data, an overwhelming 46 of the 49 contractions (94 percent) contained the apostrophe. This IM pattern is corroborated by Lauren Squires, who also analyzed contractions and apostrophes in college-student IM. Out of 218 contractions in her female IM conversations, 85 percent included the apostrophe.[53]

Punctuation of Sentences

Sentence punctuation includes a wide range of markings: capitalization; sentence-internal pauses like commas, colons, semicolons, and dashes; and sentence-final markers such as periods, question marks, exclamation marks, and sometimes ellipses. We investigated three aspects of sentence punctuation in texting and IM.

We began by looking at whether punctuation was used at the ends of messages or at the ends of sentences that weren't the final sentence in a transmission. (Remember that messages often contained more than one sentence.) Next, we examined the kind of punctuation used at the ends of sentences, specifically how often questions were marked with a physical question mark, in comparison with how frequently sentences functioning as declaratives, imperatives, or exclamations ended with periods, exclamation marks, or equivalent punctuation (ellipses, dashes, commas, or emoticons). Finally, we took a closer look at ellipses and dashes.

Seventy-one percent of all text messages and 65 percent of the IMs had no punctuation at the end of the message. If we look at all the sentences (not just the ones at the ends of the entire transmission), the numbers change somewhat: 61 percent of the sentences in text messages and 55 percent of the sentences in IM had no punctuation mark at the end.

Why the discrepancy? Because in messages containing more than one sentence, the students were selective about where they used punctuation. If we look just at sentences that appear earlier on in the message (that is, not at the end), only 46 percent of these sentences in text messages and 22 percent in IM lacked a punctuation mark at the end of the sentence.

To summarize the cases where there was *no* sentence-final punctuation:

	texting	*IM*
message-final sentences:	71%	65%
all sentences:	61%	55%
sentences *not* at the ends of messages:	46%	22%

Here's an example (from the texting data) of sentence-final punctuation used when the sentence is *not* at the end of the message, but *no* such punctuation at the end of the transmission:

sentence 1:   I mean I just want to see you . . .
sentence 2:   I'm just stressed and overwhelmed

The next analysis compared the presence or absence of question marks (for semantic questions)—such as "what's a few more years right?" with use of other marks at the ends of sentences functioning as declaratives, imperatives, or exclamations. Some examples from texting include:

declarative: Im at work til like 930
imperative: yes call me
exclamatory: omg!!!

We coded all sentences with regard to grammatical intent (not punctuation): Did the sentence ask a question or not?

In texting, 19 percent of all sentences were questions, while only 10 percent of the IM transmissions were. (All the rest of the sentences were declaratives, imperatives, or exclamations.) When we examined use of sentence-final punctuation for questions versus declaratives, imperatives, and exclamations combined, we found significantly higher use of question marks to end questions than of periods or exclamation marks to end the other sentence types. For text messaging, 73 percent of questions were ended with a final question mark, while only 30 percent of declaratives, imperatives, and exclamations (combined) bore sentence-final punctuation. In the case of IM, all of the interrogatives ended in question marks, while only 41 percent of the remaining sentences were marked by punctuation.

We were also curious to see what kinds of questions the students were asking. While 10 percent of the interrogative sentences in the IM data involved coordinating social activities, 90 percent of the texting questions did. (Think of all those AU and Michigan students who said "arranging to meet" was their main reason for doing text messaging.) At least in the case of texting, use of question marks makes pragmatic sense, because they signal the reader (like a flashing red light) that there's a message calling out for a response: coming to the party? where do U want 2 eat? lend me $20?

The other punctuation marks we analyzed were ellipses and dashes. In formal writing, ellipses stand in lieu of omitted text, such as in a quoted passage. More informally, ellipses can indicate speech trailing off ("I know what you mean. . . ."), be used for dramatic effect ("and the winner is . . . Helen Mirren"), or separate sentences in place of a more standard period (as in "It's hard to read the gambler's motives . . . he's stalling for time"). Since both text messaging and IM tend to be informal and commonly contain more than one sentence, analyzing ellipses made linguistic sense. In coding the data, we found a small number of dashes that appeared to be functioning similarly to ellipses.

In texting, there were 29 instances of ellipses and one dash, while the IM data contained 19 examples of ellipses and 6 dashes. Here are the functions they served:

Using ellipses or dashes instead of	Texting	IM	Example
Period	80%	88%	That is a strange question. . . . [IM]
Question mark	7%	0%	how come you called me so late last night. . . . [texting]
Comma	3%	12%	Soooo . . . i have been doing a lot of spying lately [IM]
Other type of pause	10%	0%	tell me . . about somethin bout joes's lol [texting]
TOTAL	100%	100%	

Lessons Learned

Texting and IM turned out to be similar in the following ways:

Feature	Texting	IM
Emoticons, lexical shortenings		
emoticons	<1% of words	<1% of words
acronyms	<1% of words	<1% of words
Sentence punctuation		
overall sentence punctuation	39% of sentences	45% of sentences
transmission-final punctuation	29% of sentences	35% of sentences
transmission-internal punctuation	54% of sentences	78% of sentences
use of required question mark	73% of questions	100% of questions
use of required period	30% of other sentences	41% of other sentences

At the same time, texting and IM also showed some clear differences:

Feature	Texting	IM
Length		
transmissions (in words)	7.7 words	6.0 words
transmissions (in characters)	35 characters	29 characters
one-word transmissions	3.7% of messages	18.8% of messages
multi-sentence transmissions	60% of messages	34% of messages
sentences per transmission	1.8 per transmission	1.3 per transmission
Emoticons, lexical shortenings		
abbreviations	3% of words	0% of words
contractions	85% of potential	68% of potential
apostrophes	32% of contractions	94% of contractions

The paucity of emoticons and acronyms in both texting and IM is precisely what Crispin Thurlow and Alex Brown found in their work on texting in the UK, and what Sali Tagliamonte and Derek Denis discovered in analyzing Canadian IM. While there aren't previous reports on sentence punctuation in texting or IM, our study indicates that usage is hardly scattershot. Students often omitted message-final marks (especially periods), but their overall punctuation choices tended to be smartly pragmatic. Greater use of punctuation in IM than in texting probably reflects the relative ease of input.

Judging from our sample, American college-student text messaging and IM differed in several interesting ways. Text messages were consistently longer and contained more sentences, probably resulting both from cost factors and the tendency for IM conversations to be chunked into sequences of short messages. Text messages contained many more abbreviations than IMs, but even the number in texting was small.

In their study of text messages written by a sample of mostly British female college students in Wales, Thurlow and Brown reported that abbreviations accounted for nearly 19 percent of the words in their data, which they judged to be a small proportion. For sake of comparison, we re-coded the abbreviations in our texting sample to more closely approximate Thurlow and Brown's scoring rubric (essentially, including all our "maybe" examples). The result: Our texting corpus still contained less than 5 percent abbreviated words.

American texting and IM data also seem to diverge in terms of contractions and apostrophes. More contractions appeared in texting, but texting used only one-third of the apostrophes found in IM. Greater use of contractions in texting probably reflects the higher tendency to use abbreviated forms (compared with IM), which in turn is expected with an awkward input device. Paucity of apostrophes in texting undoubtedly results from input complexity.

This study offers a first run at charting the language of college-student texting in America (as of the mid-2000s), along with a sense of how texting compares with IM. But what about the less-easily measured question of whether prior experience with word processing and IM is reflected in American texting?

The answer is unclear, especially in light of differences between input devices. Many Americans have little practice texting on phone keypads. Thurlow and Brown's British subjects sent text messages twice as long as American texts (14 words versus 7.7; 65 characters versus 35), which may reflect their students' more extensive texting experience. Alternatively, longer British texts may result from historically limited access to IM (a medium in which combined seriatim transmissions commonly exceed 14 words). If you have a lot to say, pack it into a text message, since IM is less of an option. Another possible explanation is that text-messaging length varies among accomplished users from one culture to the next. In Norway, for instance, text messages written by young adults average around 7 words, with 30–45 characters.[54]

How do we explain the pattern of uncontracted versus contracted forms in our texting data? The fact that 13 uncontracted forms were used (out of 85 potential contractions) might indeed reflect computer keyboard writing habits, as we saw in IM. Alternatively, the uncontracted forms could be intentionally chosen to indicate verbal emphasis—or, equally possibly, to avoid the complexity of creating apostrophes in a contracted form. In essence, the more formal-sounding expression might actually be a typing shortcut.

And what about the apostrophes that did appear in texting contractions? Laborious insertion of 23 apostrophes (out of 72 contracted words) might reflect writing habits formed at the computer keyboard (and carried over into IM). Thurlow and Brown also reported some use of apostrophes in their British-English study, so the issue may be general knowledge of punctuation rules, regardless of input medium.[55]

As with our mobile phone questionnaire, the texting study whets our appetite for more data, especially over time. What will texting look like in America in another five years? Outside the United States, will text messages be reshaped by increased use of instant messaging on computers? Or perhaps both of these technologies will be left in the dust by some new form of electronically-mediated language.

8

• • • "Whatever"

Is the Internet Destroying Language?

Shirley Jackson's "The Lottery" is an American classic. Published in the *New Yorker* in 1948, the story recounts how each June, a New England town randomly selected one of its citizens to be stoned to death to atone for everyone's sins. The story is fiction, but you can almost hear Tessie Hutchinson's screams as the rocks begin to fly. I have a particular soft spot for Tessie, having drawn her role in a high school play.

Finding scapegoats for our foibles is hardly restricted to fiction. "Why," the boss asks, "were you late to work?" Answer: "My alarm didn't go off." "Why," society asks, "is juvenile delinquency increasing?" The misogynist answers: "Because women aren't staying home with the kids."

Language is a human behavior over which people have historically made a considerable fuss. We talk about one variety of a language being the "standard," with the implication that any other version isn't as good. Language academies such as *l'Académie française* are established to ensure the citizenry knows someone is preserving distinctions between linguistic right and wrong. In the case of English, a cluster of self-appointed prescriptivists in the eighteenth and early nineteenth centuries forbade us from using double negatives or putting prepositions at the ends of sentences. As recently as the early twentieth century, Henry Watson Fowler, author of the classic *Dictionary of Modern English Usage*, literally invented distinctions between the words *masterful* and *masterly* to suit his whim.[1]

What if people fail to obey the rules? If they're in school, you can give them poor grades; if they are applying for jobs, you can pass them over. But if neither case applies, the best bet is to bemoan the situation in the media. The court of public opinion (primarily newspapers, radio, and blogs) renders judgment and sentencing.

Distinguishing between language change and language decline is very tricky business. Since yesterday's change is often today's norm, we may simply need to wait long enough before an innovation stops being treated with opprobrium by language elites. At the end of the fourteenth century, Chaucer wrote the words "hath holpen" in his Prologue to the *Canterbury Tales*, where today we write "has helped." No grammar police come after us. In a similar vein, consider the word *let*, which has two very different sets of meanings. The first is 'permit' or 'allow' or 'rent' (as in "Let me help you" or "an apartment to let"); the second is 'hinder' (as in a "let ball" in tennis—a serve that hits the net). Historically, the 'allow' meaning was paired with the Old English word *lætan*, while the 'hinder' meaning was expressed with a very different word, *lettan*. Through the vagaries of language change, *lætan* and *lettan* both became *let*. Again, the grammar police are nowhere in sight.

But what about changes or blurred distinctions where living, breathing people still swear by the "standard"? Take the words *may* and *can*, which traditionally refer (among other things) to 'permission' and 'ability,' respectively. Does it matter if an educated speaker of English says, "Can [= permission] I please have a cappuccino?" Or what about the words *capital* and *capitol*? The first refers to a city that is the seat of government of a larger jurisdiction (as in "Albany is the capital of New York"); the second denotes the physical building that houses a legislative body ("The Capitol Building is located in Washington, the nation's Capital"). Does it matter if the spellings are kept distinct—or if they go the way of *lætan* and *lettan*?

The general response of most linguists to the prospects of language change is "Bring it on!" The issues, however, are more nuanced. For the moment, suffice it to say there is active disquietude about English language standards (especially regarding writing), and that among the more normatively minded, many scapegoats have been proposed. Laid-back parenting and permissive educators have long headed the list. But there's a new candidate: electronically-mediated language. If (so the argument goes) kids are emailing and IMing and text messaging using degraded language, then it's no wonder that spellings such as *U* for *you* or *B4* for *before* are cropping up in school assignments. B4 you know it, we'll be writing verbal pabulum.

Like Tessie Hutchinson, electronic language has been singled out for stoning. To drive home the point, Crispin Thurlow analyzed more than 100 articles from the international English-language press, written between 2001 and 2005, on electronically-mediated language.[2] Scores of journalists are apparently in agreement that our linguistic prospects are bleak. A number proclaim that email, instant messaging, and text messaging have created a whole new language, apart from standard English:

> "A new language of the airwaves has been born" (*Guardian*, June 26, 2003)

"Not since man uttered his first word and clumsily held a primitive pencil nearly 10,000 years ago has there been such a revolution in language." (*Daily Post*, September 26, 2001)

That new language is degraded:

"Texting is penmanship for illiterates." (*Sunday Telegraph*, July 11, 2004)

"The English language is being beaten up, civilization is in danger of crumbling." (*Observer*, March 7, 2004)

But worst of all, electronically-mediated communication is contagious, polluting traditional writing:

"text chats are starting to bleed over into other aspects of life" (*National Post*, January 4, 2005)

"Appalled teachers are now presented with essays written not in standard English but in the compressed, minimalist language of mobile phone text messaging" (*Scotsman*, March 4, 2003)

"[T]he changes we see taking place today in the language will be a prelude to the dying use of good English" (*Sun*, April 24, 2001)[3]

Our data on IM and text messaging suggest that at least among a sample of American college students, electronic language is at most a very minor dialectal variation. Yet these findings notwithstanding, there's an international perception that computers and mobile phones are affecting everyday language, and that these effects are generally not for the better.

This chapter considers whether written language used on the Internet (and by extension, on mobile phones) is influencing offline writing, and perhaps even speech. The discussion unfolds in three stages. First, the current state of offline writing—the kind that appears, for instance, in newspapers, essays, or formal advertisements. Here is where we should look for our modern-day language culprit—if indeed one exists. The second stage addresses the normative question: Is electronically-mediated communication a linguistic free-for-all, or are there shared rules that users either follow or violate? Finally, I'll suggest several ways in which the Internet may actually be shaping the ways we write and speak.

• • • DRESS-DOWN EVERYDAY

If you stepped inside a major American law firm twenty years ago, you would have seen a Brooks Brothers version of utopia: all the associates and partners suited up in their business best, even if they were mired in brief-writing and slated to meet with no one other than their secretaries. Over time, dress-down Fridays offered some sartorial respite, even though the attire was not what most people wore for romping with the dog. Today, jeans and running shoes fill the halls of many of those same firms, and not just on Fridays.

Law-firm attire hardly evolved in a vacuum. Over the past fifty years, American society has become increasingly informal. You see it in the way chairs are arranged in classrooms—less often in serried rows and more commonly in a circle or in pods. You hear it in the way we address people we don't know or who are at different points on the traditional social hierarchy. Thirty years ago, I was "Ma'am" or "Ms. Baron" to the appliance salesman or airline ticket agent. Today, I'm "Naomi." And you see it in the way we write.

Proofreading and the Decline of Public Face

We earlier introduced Erving Goffman's work on presentation of self, according to which people portray themselves to others as if they were actors on a stage, choosing their costumes, words, and affect, to indicate their desired role. Do we want to be seen as wealthy or in need, intelligent or clueless, ambitious or cautious, leaders or followers? This public face we display to others is shaped by individual bent as well as by community norms that are in vogue. High fashion of the 1920s is today relegated to period drama, and the Boston accent that sounded so prestigious in the Kennedy era now sounds, well, simply like an educated Boston accent.

How have mainstream attitudes toward public face been changing since World War II? Three significant shifts have been a reduced emphasis on social stratification and overt attention to upward mobility; notable disconnects between educational accomplishment and financial success; and a strong emphasis on youth culture. Each of these transformations has contributed to more casual behavior in dress, language, and standards of politeness, often resulting in diminished attention to monitoring how we present ourselves to others. Obviously, many Americans still consciously work to shape and convey their images. People wear funky hats, put on accents, or spuriously report having advanced degrees. As we have seen, online communication offers boundless opportunities for defining our identities or at least presenting ourselves on "our best day." However, in the process of orchestrating public identities, contemporary Americans are less formal and less interested in overtly jockeying for social position than in times past.

How do these shifts play out in practice? As increasing portions of the American population view themselves either as middle class or legally protected to be accepted "as they are," there's less impetus to learn the fine points of etiquette or dress up for a job interview. When high school or college dropouts become multimillionaire (or billionaire) rock stars, basketball players, or computer CEOs, the public's belief that education is linked to financial success weakens. So, too, does public commitment to developing the sophisticated thought and language that higher education traditionally nurtures.

Another factor is Americans' fixation with appearing young. Youth-driven entertainment (from Disney to Eminem) and teenage lingo (think of *cool* or the pandemic use of *like*) permeate the tastes of adults who once prided themselves on being more sophisticated. Increasingly, the cultural and linguistic norms of adults, to which the nation's children formerly aspired, are being dumbed down to match popular teenage practice.[4]

One recent manifestation of this move away from putting on a well-scrubbed public face is an emerging ambivalence about individual privacy. We are troubled when companies can track our buying habits, and we fight to limit access to personal medical records. Yet many respectable citizens don't think twice about undressing before curtainless windows, revealing intimate details of their lives on blogs, or appearing on reality shows.

Lack of social inhibition is also manifest in a lot of the email being sent. One of my favorites is a message I received from a graduate student in the Philippines, requesting help on his master's thesis on computer-mediated language. The thesis, it seems, was due very soon, and his library's resources were sparse. After presenting me with a long list of questions, he closed with, "OK NAOMI. I really need your information as soon as possible." I responded (politely and briefly), though "OK NAOMI" seemed rather presumptuous from a person who was probably half my age and was, after all, seeking my help. My correspondent felt I was withholding information. He wrote back: "If you know something . . . tell me." Email is legendary for inviting personally aggressive behavior, but here was an assault against rules of social decorum.

Contemporary writing is increasingly informal. Part of that informality comes from the growing trend for writing (both online and off) to approximate informal spoken language. We have already looked at the extent to which IM bears speechlike qualities. In an earlier book (*Alphabet to Email: How Written English Evolved and Where It's Heading*), I built the larger case that traditional writing increasingly mirrors informal speech.

An additional force behind this informality is a redefinition of what is ephemeral and what is durable about linguistic expression. Recording devices aside, spoken language is inherently evanescent. Once we have uttered a sentence—grammatical or otherwise—it's gone. Part of the genius of human conversation is that we commonly ignore (or at least forgive) one another's mispronunciations or slips in grammaticality.

A further explanation may be that our goals for teaching writing have shifted. Much as the communicative approach in foreign language pedagogy has significantly supplanted grammar-driven lessons, our schools are increasingly encouraging students to write what is on their minds, with the objective of self-expression taking precedence over "proper" writing style.

For better or worse, writing used to be different. With writing, you were supposed to be on best behavior, because someone could re-read what you had inscribed. Grammatical or orthographic errors might come back to haunt you. Electronically-mediated communication is written, though we tend to think of it as more like transient speech. Most messages we read and delete; like speech, they're gone. But in a sense, we increasingly do the same thing with print, and not just the newspaper we discard at the end of the day. Given the explosion of printed material—from the pile of junk mail that arrives daily to the pulp fiction we leave on the plane to books we buy but never read, the tangible written word is something over which we, paradoxically, seldom linger. As a result, our standards as readers, writers, and even publishers are becoming less discerning.

Take what the British call the "greengrocer's apostrophe," named for aberrant signs advertising *cauliflower's* or *carrot's* in local fruit and vegetable shops.[5] Its American counterpart: a sign on a locksmith shop in Snowmass, Colorado, informing its patrons "FOR EMERGENCY'S CALL." Would you give this locksmith your business?

More chilling examples come not from the world of tradesmen but from literary professionals. In recent years, respected publishing houses have inexplicably mangled texts I submitted on computer disk. My allusion to Homer (of *Iliad* and *Odyssey* fame) became "home run." *Albany* morphed into "AlbaN.Y." The grand prize, though, goes to a venerable press chartered by England's Henry VIII in 1534.

The "Whatever" Theory of Language

Since 1857, the *Atlantic Monthly* has brought the fruits of highbrow fact and fiction to a select audience of cultured Americans. Almost three hundred years earlier, an even older publishing institution, Cambridge University

Press, issued its first book. On page 97 of the December 2002 issue of the *Atlantic*, Cambridge University Press ran a full-page advertisement, inviting readers to "PAMPER YOUR INTELLECT," recommending titles that might serve as mental spas. But instead of a pampering, the discerning reader was dealt an unkind blow. The ad contained no fewer than six proofreading errors: proper names were misspelled ("Immanuel Kant" became "Immanuel Kent"), common nouns were treated no less kindly (*Environment* degenerated into *Enviorment*), and subtitles on book jackets did not always match up with those listed in the text ("My Father's Life with Bipolar Disorder" morphed into "My Father's Life with Bipolar Disease").

Did Cambridge simply have a bad day, or is editorial sloppiness becoming ubiquitous? Casual observation suggests the latter. Look no farther than official signage for the Washington, DC metro system. The name of my university's subway stop appears two different ways: sometimes correctly as "Tenleytown" (one word), yet other times with an incorrect two-word version, "Tenley Town." The issue, I suggest, is not just a devolution of proofreading skills but a quiet revolution in social attitudes toward linguistic consistency.

Language as Rule-Governed Behavior

For more than a century, practically all linguistic theorists have shared the assumption that human language (including change in language over time) is governed by laws, consistent patterns, or rules. The late-nineteenth-century Neogrammarians and early-twentieth-century European and American Structuralists all spoke of linguistic regularity. The best known contemporary model is Noam Chomsky's, which characterizes language (at all levels: phonological, morphological, and syntactic) as rule-governed.[6] Those rules constitute the bedrock upon which the linguistic competence (or "knowledge") of individual language users is based.

It doesn't take a professional linguist to recognize that languages facilitate communication precisely because they can be described in terms of principles shared by members of a speech community. Agreed-upon interpretations of words or phrases make it possible for people to use language (rather than, say, charades or brute force) to get their meanings across. If I say "chocolate mousse" when I mean "strawberry shortcake," it's no surprise when I fail to get the dessert of my choice. And if I declare "Suzette should cut Manfred's hair" when I mean for Manfred to shear Suzette's locks, the duo can hardly be expected to comply with my intent. Linguistic communication is generally successful because we understand one another's pronunciation or handwriting, agree what words are referring to, and share our comprehension of grammatical relationships in the sentences we speak and write.

In talking about how rules define language, linguists sharply distinguish between descriptive rules (what people in a speech community "know" about their language by virtue of growing up within that milieu) and prescriptive rules of the kind we described earlier in this chapter, such as no sentence-final prepositions or distinguishing between the words *can* and *may*. The realm of written language is overwhelmingly the domain of prescriptivism, which dictates how words are spelled, how punctuation is distributed, and what special formalities you need to follow, including proper salutations in letters and clear transitions between paragraphs.

Linguistics concerns itself with descriptive rules, dismissing prescriptivism as an artificial phenomenon having more to do with social class than with speakers' "natural" knowledge of language. Since the majority of linguists study spoken, not written language, the state of prescriptive judgments in writing hasn't been on the radar screens of most practitioners in the profession.

Breaking the Rules

We all know that language rules are routinely violated. Edward Sapir reminded us back in 1921 that however useful the notion of a grammar, "all grammars leak"—that is, there are always exceptions to the regularities we write rules to characterize.[7] The way to form a plural noun in English is to add some version of an "s" sound, as in *pencil/pencils*. But then what about *child/children* or *man/men*?

Even if speakers "know" both the rules and the exceptions, they make mistakes—what Chomsky would refer to as performance errors. Another sort of rule violation appears when language is in the process of changing. For example, many older, educated speakers of American English still find jarring the number agreements in sentences such as "**Volvo** continues **their** commitment to safety" (heard on National Public Radio) or "**None** [of the former prisoners] **were** in court yesterday" (from a news story in the *New York Times*).[8] Given the growing tendency for speakers to view many singular subjects as if they were plural, we can reasonably predict that sometime in the not-too-distant future, English will "regularly" treat names of companies (such as Volvo) and quantifiers such as *none* as plural, in part to avoid the social indelicacy of having to specify gender in an antecedent pronoun: Is it "No one raised **his** hand?" "No one raised **her** hand?" Why not simply avoid the problem with "No one raised **their** hand?" (Before you protest, remember that this is the same kind of linguistic process that eliminated Chaucer's "hath holpen.")

Exceptions to language rules have another possible source: language-users simply not knowing which spoken (or written) pattern conforms to the rules. In speech, we often dismiss some of these confusions as fall-out from pre-scriptivism or incomplete language change. Should we say "Who do you trust?" or "Whom do you trust"? (This controversy has actually festered for centuries.) In writing, there's the gaff in Bill Gates's 1995 book *The Road Ahead*, in which Gates (or his co-authors) could not keep straight when to use *affect* and when *effect*.[9]

But the problem is more pervasive than indecision on a couple of word choices. Instead, there is another possibility: We are raising a generation of language users (who, in turn, impact the linguistic patterns of their elders) that genuinely does not care about a whole range of language rules. Whether the issue is spelling or punctuation, verb agreement or pronoun choice, there seems to be a growing sense of laissez-faire when it comes to linguistic consistency. It used to be that when I walked into a room and mentioned I was a linguist, people nervously declared, "I'd better watch my grammar." Those days are long over. Instead, students increasingly look askance when I painstakingly correct their linguistic faux pas. They ask, "What's the big deal?"

What is going on?

The *"Whatever"* Generation

A convergence of forces is engendering a new attitude toward both speech and writing. We might dub this attitude "linguistic whateverism." Its primary manifestation is a marked indifference to the need for consistency in linguistic usage. At issue is not whether to say *who* or *whom*, or whether *none* as the subject of a sentence takes a singular or plural verb, but whether it really matters which form you use. This challenge to the fundamental principle of language as rule-governed behavior is less a display of linguistic defiance than a natural reflection of changing educational policies, shifts in social agendas, a move in academia toward philosophical relativism, and a commitment to life on the clock.

First, education. As children, we naturally absorb the spoken language of our environs. Those born in Mississippi generally sound like Mississippians, New Yorkers like New Yorkers. Schools then further shape the way we speak—and write. Normative instruction has traditionally instilled an awareness that rules exist for using language properly.

Not surprisingly, a laissez-faire approach to language instruction begets a more casual attitude toward linguistic consistency. Since World War II, education in America has become increasingly informal, student-centered, and non-normative.[10] As faculty, we have been trained to celebrate what our students have to say and encourage them to express themselves—literally in their own words. Is it therefore surprising that students take us at our word when we de-emphasize fine points of grammar and spelling?

Next, the shifting social agenda. The United States has long represented itself as welcoming immigrants and tolerating divergent peoples and customs. Whatever the historical follow-through on these claims, recent decades have witnessed a concerted move toward multiculturalism, with the result that national rhetoric (and curricular design) reflect a legally and pedagogically structured celebration of individual and group differences. I'm not suggesting we devalue the importance of teaching diversity—which is an essential social and educational goal. Rather, we need to recognize that by its very nature, diversity is at odds with a normative social model.

The diversity agenda has linguistic implications. We teach our children not to pass judgment on regional dialects or non-native speakers. In the process, we loosen our grip—for better or worse—on a notion of linguistic correctness and consistency.

A related factor in the recent decline in linguistic norms has its roots in contemporary patterns of academic discourse. Many of us in the academy have noticed that the present generation of university students is more reticent than their predecessors to engage in debate or to criticize the words or actions of others.[11] When challenged to make judgments or take sides, a common refrain is "Whatever." Anything's OK. Let's not fight over it. Whatever you do or say—including how you say or write it—is fine.

And finally, current writing patterns reflect the haste with which "finished" writing is now often produced. As modern Western society increasingly began living "on the clock," we sped up many aspects of daily activity (fast food, express check-out lanes in the grocery store), including the way we write. Mechanical devices (first the typewriter, then the stand-alone word processor, now the networked computer) have made it increasingly easy to prepare and dispatch documents, often multitasking as we go, without pausing to reflect upon whether we have actually written (and sent) the text we intended. In the next chapter, we explore in some detail how writing went from being a contemplative activity to a rushed job.

Language in the New Millennium

No one (including me) is suggesting that the English language is about to disintegrate into a babel of unstable idiolects. The changing attitudes toward language consistency we've been describing are still largely at the edges of linguistic regularity, particularly involving writing mechanics and the familiar uncertainty zone containing the likes of *who* versus *whom*, *affect* versus *effect*, and *it's* versus *its*, with which native speakers have been wrestling for some time. What is new—and significant—is the emerging attitude of the "whatever" generation that sustained concern about linguistic consistency may be foolish (and perhaps even, to borrow from Emerson, the hobgoblin of little minds).[12]

What are the potential ramifications of this attitude for the shape of English usage in the coming decades? Some plausible scenarios:

- Writing will increasingly become an instrument for recording informal speech rather than the distinct form of linguistic representation that emerged by the end of the seventeenth century in England.[13]
- As a literate society, we will continue to write, but will revert to an attitude toward spelling and punctuation conventions redolent of the quasi-anarchy of medieval and even Renaissance England. In Middle English, for example, the word *nice* (which then meant 'ignorant' or 'foolish') might be spelled "nis," "nys," "nice," or "nyce." Later, William Shakespeare spelled his own name at least six different ways. And commas, colons, and capitalization were sometimes anyone's guess.
- We will see a diminution in the role of writing as a medium for clarifying thought (see chapter 9).
- Language (both spoken and written) will play a reduced role as a social status marker. If Lyndon Johnson, with his genuine Texas drawl, were president of the United States today, he would take much less flak for his accent than he did in the 1960s. Or, speaking of the adoptive Texan George W. Bush, the humorist Dave Barry wrote, "If you have good language skills, you will be respected and admired; whereas if you clearly have no clue about grammar or vocabulary, you could become president of the United States. The choice is yours!"[14]

Modern linguistic theory eschews passing judgment on any linguistic variant, and I am not about to do so now. Rather, I'm suggesting that should linguistic entropy snowball, we may discover that personally expressive, culturally accommodating, and clock-driven language users will find it increasingly difficult to understand one another's nuances.

• • •

You'll notice that almost none of our discussion of dress-down language or the "whatever" approach to speech or writing invoked computer-based technology. Computers are not the cause of contemporary language attitudes and practices but, like signal boosters, they magnify ongoing trends. If a sprinter can do the 100 meters in 12.3 seconds wearing old-fashioned sneakers, think what a pair of high-performance running shoes might do for his time. If I tend to be a bad speller to begin with, brace yourself for my communiqués when I leave the safety net of spell-check in word processing and move to IM or texting.

Yet when it comes to electronically-mediated language, there's an additional set of variables at work. These involve the broad issue of writing style. Long before computers came along, societies evolved complex arrays of language styles to fit different circumstances. Just as you wouldn't wear a bathing suit for an audience with the pope, you would sound odd indeed if you addressed a New York City pretzel vendor with "Excuse me, but might I trouble you for two pretzels" or said to a potential employer, "C'mon! Gimme a job!" As language users embedded within social communities, part of what we learn as speakers (and writers) is how and when to say (or write) what to whom. Notes passed in class are not as formal as job applications. Cheers acceptable at a football game are out of place at the opera.

Because electronically-mediated language is fairly new, its users are still in the process of settling upon conventions that ostensibly will become the new rules to be followed or broken. Novelty of the technology is a large part of our problem in deciding where we stand on whether email, IM, and texting are ruining language.

• • • RULES AND AUTHORITIES

In 1999, Constance Hale and Jessie Scanlon published their revised edition of *Wired Style*. While other etiquette volumes, both before and since, have approached online writing with an eye toward business users, Hale and Scanlon had a more laid-back audience in mind. The authors pointedly scoffed at the idea that email should be subject to editing—either by sender or receiver. Some samples:

> "Think blunt bursts and sentence fragments.... Spelling and punctuation are loose and playful. (No one reads email with red pen in hand.)"

> "Celebrate subjectivity."

> "Write the way people talk. Don't insist on 'standard' English."

> "Play with grammar and syntax. Appreciate unruliness."[15]

The authors propose something of a flower-child approach to email. But seen in perspective, they have as much potential authority over what email style should look like as self-proclaimed eighteenth- and nineteenth-century prescriptivists such as Bishop Robert Lowth had over the structure of English: Declare yourself an authority, and see if anyone follows.

Are there rules for the likes of email, IM, and texting? If so, who sets them?

Who Sets Online and Mobile Style?

Can you divorce your spouse by email? That question was handed to a court in the United Arab Emirates, when an American of Arab descent emailed his Saudi wife to break their marriage bonds. Islamic law holds that a man may divorce his wife

> simply by telling her "I divorce you," if certain conditions are met. [The *Gulf News*, which reported the story,] said the court would have to rule if the notification of divorce through the internet was valid under the Gulf Arab emirate's laws, or whether it should have been delivered verbally.[16]

What else should you do—or not do—by email or IM or text messaging? Send an apology? Your condolences? News to your former spouse of your remarriage? Our issue is not so much with answers to these queries but with how we arrive at resolutions.

The rise of new technologies, or popular access to them, often leads to new rules. With the coming of the telegraph, both legal and social precedents needed to be re-examined. Can you get married by telegraph? By the 1870s, people did.[17] Are the contents of telegrams privileged information, as is the case with letters?[18] To settle such questions, treatises devoted to telegraph law began to appear in the later nineteenth century.[19]

The telegraph also presented challenges to those creating messages: What should they say? Etiquette guides offered sample wording for commonly sent telegrams, such as reserving a hotel room in another city. Western Union supplied "prepared texts for those who needed help in finding the right words for the right occasion."[20]

The telephone engendered a similar need for working out appropriate linguistic conventions, beginning with the simple challenge of a conversational opener. When Alexander Graham Bell first marketed the telephone, he recommended saying "Ahoy" (normally used for hailing ships) to begin a phone call. Because telephone lines of the day were always "live," individuals placing calls needed some means of signaling their presence and desire to talk with the person at the other end.

To appreciate what happened next, you need to be aware of the bitter rivalry between Bell and Thomas Alva Edison. Edison, who had vied with Bell over first patent rights on a talking telephone, threw his inventive talents into an alternative technology—the phonograph—designed to capture spoken acoustic signals onto a waxed cylinder that could then physically be delivered to the intended recipient.[21] In the summer of 1877, while experimenting with his early phonograph, Edison shouted "Halloo" (a traditional hunting call to the hounds) into the mouthpiece of the device.

Two months later, in corresponding with a friend about how to solve the signaling problem on telephone calls, Edison suggested a variant of "Halloo." Edison wrote: "I don't think we shall need a call bell[,] as Hello! can be heard 10 to 20 feet away." Conveniently, Edison had been hired by Western Union to design a telephone unit that would compete with Bell's. As a result, Edison was in an excellent position to lobby for use of "Hello" (against his rival's "Ahoy") as an opening telephone greeting.[22]

But there was a problem. At the time, "Hello" (and its variants) were viewed as vulgar language. Etiquette books inveighed against Edison's greeting. Bell's company, AT&T, fought to suppress use of "Hello," and as recently as the 1940s social arbiter Millicent Fenwick deemed the word acceptable only under limited circumstances.[23] But such prescriptivist efforts were of no avail. "Hello" had moved into common usage by the turn of the twentieth century. Over time, it became not only the standard way to begin a phone conversation but (at least in the United States) the normal greeting for initiating face-to-face encounters. To the modern ear, "Hello" comes across as a formal, polite greeting—particularly in comparison with "Hi" or "Hey."

Much as Edison was personally responsible for the way we open modern telephone calls, online-writing pundits follow either common sense or personal whim. Etiquette maven Judith Martin (aka Miss Manners) declares matter-of-factly that

> serious presents, such as wedding presents, and serious hospitality, such
> as an overnight stay, require serious letters where the seriousness of paper
> and ink may be experienced first-hand—in other words, by mail, rather
> than fax. . . . E-mail . . . is excellent for spontaneous little bursts of gratitude
> and a supplement to big ones.[24]

On what authority does Martin legislate these practices? Her own. Virginia Shea, author of *Netiquette*, acknowledges that she made up many parts of her book as she went along.[25] David Shipley and Will Schwalbe's delightful guide *Send* offers sound advice for those using email either professionally or to make a good personal impression, but their suggestions come from their heads and hearts, not with the imprimatur of anyone other than their publisher.

Should Violators Be Prosecuted?

We've seen that usage and style issues in language come in two varieties. There are the traditional rules for grammar, punctuation, and spelling. And then there are questions of what kind of language (or language medium) is appropriate for a particular kind of message. Must condolences still be expressed in a handwritten letter or will an email (or e-card) do?[26] When should you write out "by the way" rather than *btw*?

Contemporary hand-wringing over electronically-mediated language targets all of the above, but we need to stop laying the blame for contemporary attitudes regarding linguistic correctness at the feet of IM or texting. The "whatever" attitude got there first. What's more, under many circumstances, choice of medium and tone is ultimately the private concern of writers and their audience. Whose business is it if I send a text message to a friend that leaves apostrophes out of contractions or an IM containing a misspelled word? If the recipient doesn't care, why should anyone else?

We also need to give credit where credit is due. By the time most youth are in middle school, they have a very clear understanding that different written styles (just like different spoken styles) are appropriate for particular settings.[27] If students seem not to know that difference, it's our obligation as parents and teachers to educate them, much as we taught them as preschoolers to say "please" and "thank you."

Anecdotal evidence suggests that a number of teachers, not wanting to be branded as troglodytes, are admitting IM novelties into classroom written assignments. No harm is done if—and this is the crucial caveat—those same teachers are ensuring their charges develop a solid grasp of traditional writing conventions as well. Unless we are willing to welcome people spelling their names six ways or using punctuation marks in free variation, we owe it to our children to make certain they understand the difference between creativity and normative language use.

We also tend to blow out of proportion the scope of IM or texting language at issue. In reality, there are relatively few linguistic novelties specific to electronically-mediated language that seem to have staying power. Abbreviations in texting? Sure. A handful of new acronyms (such as *lol* or *brb*)? Indeed. Hundreds more available? Yes, though they are unintelligible to the vast majority of even young Internet users. The idea that everyone under the age of twenty-five knows an entire new language is simply poppycock. Young

American teenage girls probably specialize in linguistic innovation more than anyone else, but most outgrow the urge when they turn their attention to actually getting their meaning across. Besides, since Americans are proficient typists, extended IMs are not really a problem, if we choose to send them. As for mobile phones, after years of practice on phone keypads, Europeans and Asians have established that constructing long text messages is no big deal. Some young people really do fill up all 160 characters without seeming the worse for wear.

Underlying the contemporary hand-wringing is actually a deeper concern: that Internet language is corrupting the way we craft traditional writing or even speak face-to-face. The time has finally come to address this question head-on.

• • • WILL THE REAL INTERNET EFFECTS PLEASE STAND UP

If this were the year 1750 and I, a gentleman, were writing to a friend, I might well close the letter with the words "Yr Hm Ser," that is, "Your Humble Servant." There was nothing particularly informal about the abbreviation; its use was simply current custom.

With abbreviations and acronyms, language users are continually re-inventing the wheel. As children penning letters to Grandma, we used to close with "XXX and OOO" (hugs and kisses), which now pop up in teenage IMs. Abbreviating *with* as *w/* may look like a novel shortcut, but Tiros, the secretary of Marcus Tullius Cicero, had the idea earlier, when he used the letter "c" as shorthand for Latin *cum* (meaning 'with'). In the same vein, abbreviations such as *U* for *you*, *R* for *are*, *4* for *four*, and obviously *yr* for *your* were around decades before they began infiltrating American text messages. Are these creations "new" in the minds of their modern users? Perhaps, but then printing was "new" to Gutenberg, presumably unaware of his Chinese predecessors.

Other obvious pieces of online lingo are original, such as *pos* ('parent over shoulder') and *lol* ('laughing out loud'). And yes, I have had college students use *lol* (along with *btw* and *b/c*) in term papers, much to my consternation—and their subsequent regret.

Are these lexical shortenings just the tip of the iceberg? In 2005, I put that question to a group of experts who were joining me on a panel I was moderating for a session on "Language on the Internet" at the annual meetings of the American Association for the Advancement of Science. In our planning, we all thought about the issue for several months but collectively came up nearly dry. After two more years of thinking, I've fleshed out the list a bit, largely by expanding its scope to include meta-issues that go beyond the usual candidates. The inventory begins with influences of electronically-mediated communication on written language, has a few things to say about speech, and then hones in on one of this book's main themes: controlling the volume of communication.

Impact of Internet Language on Writing

The most obvious influence of Internet language on traditional writing is to reinforce two ongoing tendencies we have already described. One of these is strengthening the role of writing as a representation of informal spoken language. While instant messaging is not the same as speech, IM (along with email and texting) does tend to be an informal mode of communication. As such, messages may be peppered with words like *cuz* (*because*) and *ya* (*you*), where the orthography is a representation of speech, not special electronic language. In the same vein, abbreviated forms such as *b/c*, *U*, or *R* accentuate the casual tone.

A less obvious effect of the Internet on writing is the increasing uncertainty that many of us feel about whether a clump of language should be one word, a hyphenated word, or two words. The traditional rule is deceptively simple: If a combination of an adjective plus noun comes to function as a single concept, and if in speech we put the stress on the first word (rather than equal stress on both words), then the combination should be written as one word. A standard example is *blackboard* (the expanse you write on with chalk) as opposed to *black board* (any sort of board, say the one on the floor in the corner, which just so happens to be black). Similarly, *overtime* versus *over time* or *blueberry* versus *blue berry*.

However, the story quickly becomes more complex. Take the invention of putting loose tea leaves into a small, permeable bag. What do you call the bag? The *Oxford English Dictionary* notes the phrase *tea bag* appearing in 1898. By 1936, the word was hyphenated (*tea-bag*), and by 1977 had made its way to the orthographic compound *teabag*.[28] In the computer world, we find the same process at work, though practitioners (and their publishers) often disagree whether the word is *e-mail* or *email*, *on-line* or *online*. (Curiously, the move from *off-line* to *offline* seems to lag behind its antonym.)

Or take *newspaper*, *news-paper*, or *news paper*. All three versions pass spell-check (which, for my students, is the final arbiter of orthography). What's more, at least in casual speech, all three versions sound the same—returning us to the fact that writing is increasingly representing informal speech rather than standing as a distinct linguistic form of representation. As some of my students tell me, worrying over hyphenation and compounding is so twentieth century.

How should we decide when to take the leap from hyphenation to compound? Historically, we've relied both on changing stress patterns in speech and frequency of use. The more common the term in everyday language, the faster it becomes a compound. Hale and Scanlon suggest that personal taste is a sufficient desideratum. In announcing their decision to spell the term *homepage* as one word (rather than as two or with a hyphen), they muse that "the two syllables combine to form one idea." And besides, they "like the frontier echoes of 'homestead.'"[29] Spelling change by personal decree—an interesting approach to language standards.

The Internet may, in fact, nudge language toward more compounding, though not because of a frontier spirit. Think of the spelling of URLs that we are constantly typing in when we do web searches. "Washington Boat Show" becomes "washingtonboatshow," looking for all the world like the *scriptio continua* from classical Latin, which did not place physical spaces between words. The problem is further magnified by a trend in the 1990s for commercial enterprises to compress names into single words—often in abbreviated form—using what some typographers call CamelCase. The pharmaceutical company Glaxo Smith Kline became GlaxoSmithKline, while National Westminster Bank morphed into NatWest. No wonder word division has become problematic. Increasingly, I find students are genuinely confused about when to use a compound—even with words they probably could have spelled perfectly well ten years earlier: Is it *a part* or *apart*? (The terms mean very different things, thank you. The decision is not random.)

A third potential influence of the Internet on language is computer software that's too smart for our own good. Take the case of spell-check. Say that I am uncertain whether the word *capitalization* is spelling *capitalization* or *capitolization*. In the really old days, I would be driven to look up the word in the dictionary, and perhaps, in the process, learn how to spell it. In the not-so-old days, spell-check would tell me I had a problem, but wait for me to request the correct spelling (during which time, perhaps, some learning might take place on my part). Today, however, spell-check doesn't dawdle. By the time I've pressed the space bar after *-tion*, my trusty spelling valet has already corrected the wayward "o" for me. Handy, yes, but I've missed out on a potential opportunity to improve my orthography.

Consider use of apostrophes. We are all familiar with the common confusion between *its* and *it's*. Orally, the words are homophones, so speakers are on safe ground. In writing, though, the words are distinct—that is, they should be. Some of my undergraduates tell me that when writing papers on a computer, they no longer bother to type apostrophes in contractions, since spell-check will automatically insert the punctuation mark for them. (Alas, the strategy breaks down for strings of letters such as "cant" or "wont".) The problem becomes exacerbated through proliferation of apostropheless URLs or email addresses. For example, for his column in the *Examiner*, Jonah Goldberg (editor-at-large of the *National Review Online*) uses the email contact JonahsColumn@aol.com.

Equally subversive to normative sentence mechanics such as spelling and punctuation are modern computer search engines. Suppose I'm looking for information on evolution. But it's late at night, or I'm feeling lazy, or I'm not sure of my spelling, so I type "cHarlz dARwon" into the search box. Google politely inquires, "Did you mean charles darwin?" Sure, Google, that's exactly what I meant (give or take some capitalization). Thank you for obviating the need for me to express myself clearly. The problem is that if I come to rely on Google to figure out what I meant to write, what is my motivation for expressing myself precisely in the first place?

Impact of Internet Language on Speech

Neologisms enter spoken language all the time. Some examples from the past century include *moonlighting* (with the meaning 'work a second job'), debuting in 1957, or *scofflaw* (originally referring to someone who scoffed at the law against drinking—the word was coined for a contest in early 1924, during Prohibition).[30]

New entries are sometimes carryovers from written language, especially written acronyms. Familiar examples include *RSVP* (with each letter bearing its own name), *AWOL* (sometimes rendered in speech as "a-wall"), and *ASAP* (commonly "a-sap"). Occasionally, written abbreviations make their way into speech. In the town of Storrs, the University of Connecticut is fondly known as "U-Conn." And after years of scanning parking signs to determine whether meter fees are in effect on weekends on a particular street, I now regularly inform my family, "No, we only need money Mon through Fri."

Are abbreviations or acronyms from electronically-mediated communication appearing in spoken language? Perhaps a few, though you can literally count the candidates on the fingers of one hand. Most likely to succeed? *Brb* ('be right back'). Not only do I hear it from a number of my students, but some in their parents' or even grandparents' generation have started using the term as a whimsical declaration they are not yet over the hill.

A second contender is *lol* ('laughing out loud'). True, I've not observed the letters uttered by anyone over the age of thirty, and its spoken popularity may be highly regional (still the rage at North Fork High, but entirely uncool at South Fork). Speaking of popularity, *rotfl* (as in 'rolling on the floor laughing') had a brief spoken run among the middle-school set in my neighborhood in the late 1990s ("Jake was making these crazy faces and I was like R-O-T-F-L-ing"). Lately, though, I get blank stares when I ask my students if they ever use this acronym in their speech.

Innovation doesn't always lead to staying power. *WMD* ('weapons of mass destruction') was on everyone lips in 2003, but the acronym is no longer instantaneously grasped. *Pos* ('parent over shoulder') was extremely handy five years ago when teenagers living at home wanted to keep their IMs private (away from Mom or Dad, who had just walked into the room). But if text messaging on mobile phones encroaches on the electronic turf of IM, then *pos* could become passé.

Finally, some abbreviations appearing in electronic language are really representations of spoken abbreviations and not unique to online or mobile communication. Examples include *cig* for *cigarette* or *ya* for *you*.

Impact of the Internet and Mobile Phones on "Volume" Control

Is the Internet destroying language? If you look at the effects—direct or otherwise—on traditional language, the case is highly tenuous. True, electronically-mediated language and the likes of spell-check and Google make it easy to drift into sloppy writing habits. The culprit, however, is not the technology. Depending upon how you view the situation, fault lies either in ourselves or in the more global "whatever" attitude regarding regularity in language. When it comes to speech, the potential effects of the Internet (at least as of now) are negligible at best.

Yet there is another area of interpersonal communication in which the Internet and electronic language are having a profound effect. That area is "volume" control. In chapter 3, we argued that online and mobile language technologies have increasingly made each of us a language czar. In the ensuing chapters, we documented how we block incoming IMs, package ourselves on Facebook, write vigilante blogs, use ring tones or caller ID to screen incoming calls on our mobile phones, or beep acquaintances without having to pay a connect charge. These are subtle effects, not measurable by a high school teacher's red pen or by dictionaries of new words. But these, I suggest, are likely to be among the most lasting influences that contemporary information technology has upon the ways we communicate with one another.

• • • LIVING IN AN ONLINE AND MOBILE WORLD

The subtitle of this book is "Language in an Online and Mobile World." In the course of our journey, we have explored how computer-based technology and mobile phones offer up new ways of interacting with others, increasingly on our own terms. There are, however, two larger questions we still need to address.

The first probes the impact of online and mobile language on our conception of what it means to live in a written culture. Do we now read differently? Write differently? Do we think differently as a result of our experience with networked computers?. These are the questions addressed in chapter 9.

Chapter 10 widens our lens still farther, looking at the ways in which computer-mediated language and mobile phones are affecting us as individual and social beings. The question: What are the consequences of being linguistically always on?

9

● ● ● Gresham's Ghost

Challenges to Written Culture

He was a savvy merchant and financier, counselor to royalty, but not above maneuvering for personal gain. Sir Thomas Gresham, founder of the British Royal Exchange and of Gresham College, is best known for the advice he offered Queen Elizabeth I upon her ascension to the throne in 1558. "The good and bad coin," he wrote, "cannot circulate together."[1] That is, good coinage—coins that had been struck with the proper amount of precious metals and had not been debased by users scraping off shavings—would be hoarded.

Bad money (as we say today) drives good money out of circulation. People would tend to circulate only the "bad," debased coins, even though both bore the same face value. Gresham's advice to Elizabeth was timely, since her predecessors, Henry VIII and Edward VI, had been notorious for reducing the amount of silver in English coinage, with unfortunate financial results. "Good" coins were either hoarded or sent abroad for international trade, while "bad" coins were all that was left for doing commerce at home.

For "coin" substitute "writing," and replace the word "economy" with "written culture." The growth of online and mobile technologies has fostered a steady increase in the amount of writing we do. Where previously we spoke face-to-face or picked up the (landline) telephone, we now commonly write email, instant messages, or text messages. We air our thoughts and knowledge on listservs, in blogs, on Facebook, and on Wikipedia. The sheer amount of writing we are churning out is staggering. Is this writing explosion helping drive "good" writing out of circulation?

Analogies and metaphors are not meant to imply that the two phenomena being compared are identical. This chapter is called "Gresham's Ghost," not "Gresham's Law." The issue with writing is not that people are hoarding excellent prose or only shipping it abroad. Instead, the outpouring of text fostered by information communication technology may be redefining (some would say debasing) our standards for the written word. If "good" and "bad" writing bear the same face value, motivation for struggling to produce "good" prose diminishes.

For roughly the past 300 years, the English-speaking world has functioned in terms of what has been called a written or print culture. (We'll use these terms interchangeably.) Expanded use of the printing press, a rise in literacy rates (along with growing social mobility), the spread of Bible-reading through Protestantism, and increasing linguistic distinctions between written and spoken language were just some of the forces contributing to the establishment of print culture.

Societies change with time, and so do attitudes toward language. This chapter explores major assumptions that have historically defined our notion of written culture, along with contemporary challenges to those suppositions. In particular, we look at five dimensions of the written word in which contemporary attitudes may be redefining traditional notions of reading and writing. The issue is what these changes bode for the future of print culture.

• • • PROFILE OF WRITTEN CULTURE

A favorite cartoon of mine by William Hamilton drolly illustrates the notion of a written culture, as understood by educated twentieth-century Americans.[2]

I HAVEN'T ACTUALLY BEEN PUBLISHED OR PRODUCED YET BUT I HAVE HAD SOME THINGS PROFESSIONALLY TYPED.

William Hamilton. Courtesy of the artist.

Having your work appear in print has long been a measure of success in a written culture. Being "professionally typed" is hardly as good, but at least might be valued more—so the man hopes—than handwritten scrawl.

To speak of a society having a written culture is different from observing that some people can read and write. Historically, it is not uncommon for societies with sophisticated written works essentially to function as oral cultures. In Classical Greece, literacy played an incalculably important role in the emergence of philosophical thinking. Yet fifth-century Athens retained an oral culture. Political and legal proceedings were overwhelmingly oral, and literature (the *Iliad*, the *Odyssey*, the works of playwrights and poets) was intended to be rendered aloud, not studied as written text.[3]

Looking westward, England was largely an oral culture through the sixteenth century. Wills were recorded, but until the seventeenth century did not have independent legal standing apart from the oral testimony of those who had witnessed them.[4] While medieval literacy was important in the lives of the clergy and the new Anglo-Norman nobility, the number of people who could read or write was small.[5] Moreover, even those who were literate often heard texts rather than read them. In the words of J. A. Burrow, "People in the Middle Ages treated books rather as musical scores are treated today. The normal thing to do with a written literary text...was to perform it, by reading or chanting it aloud."[6]

The oral character of most literature persisted into the time of Queen Elizabeth I and the Globe Theatre. Shakespeare wanted his poetry printed but is generally seen as having less interest in publishing his plays. Though quarto editions of individual plays appeared during Shakespeare's lifetime, the first folio compilation (which was specifically meant to be read) was done posthumously. Shakespeare largely composed his plays to be seen and, most importantly, heard.[7] The Shakespearean stage used few props, no scenery, no costumes. To understand a performance, the audience relied on listening—a skill in which they were well practiced from experience in church, Parliament, court, and taverns.

Development of a solidly written culture in the West was made possible by numerous social and technological transformations, the most important of which was printing. Although Gutenberg's Mainz Bible appeared in 1455, it took at least 200 years before print technology was generally accepted as a substitute for manuscript production and before there was a substantial audience for print.[8]

Why We Write

Why do people write things down? The reasons are many: professional, social, and personal.

Professional writing covers a multitude of functions. The oldest is administrative, evidenced by the use of Linear B for record keeping in Mycenaean Greece, around the fourteenth century BCE.[9] Another professional use of writing is for commercial purposes. Samuel Johnson famously declared that only blockheads don't write for money, though in the early days of printing, courtiers and gentlemen typically eschewed publishing their poems to distinguish themselves from the new breed of poets seeking financial gain through print.[10] More recently, professional writing has become a form of hurdle jumping, where students must churn out research papers and university faculty must publish to get tenure.

Writing also fills social functions. Since the days of early modern Europe, members of the literate class have exchanged letters, poems, and stories.[11] As literacy rates grew, new writers sometimes relied on "complete letter-writers" to provide templates for all occasions.[12] Over the past century, pre-written greeting cards became a billion-dollar industry.

And we write for personal reasons. We take notes at meetings and jot reminders to ourselves. More profoundly, some keep diaries, write poetry, compile commonplace books containing quotations from other people's writings, or publish "for the record" works others have composed. The political analyst Daniel Ellsberg exemplified for-the-record publishing when, in 1971, he provided the *New York Times* with a 7,000-page secret RAND Corporation report. That document, which came to be known as *The Pentagon Papers*, revealed hitherto unpublished information regarding America's involvement in Vietnam. Ellsberg's motivation? To stop the war.

Attributes of a Written Culture

For a written culture to emerge, a sizeable number of its members need ways of creating, disseminating, and deciphering the written word. People must have access to the tools of production (be they personal computers or quills on parchment) and knowledge of how to use them. A particularly complex writing system may limit the number of individuals having the opportunity to become literate (as happened in imperial China) or the amount of material that can conveniently be printed (the case for Japanese before the development of word processing with *kanji* and *kana* in the late 1970s[13]). William Harris argues that one reason Classical Greece remained an oral society, despite the critical role literacy played in its intellectual accomplishments, was that it lacked efficient means of duplicating and disseminating texts. In the English-speaking world, not until the development of affordable and reliable postal systems did personal letter-writing become part of the general culture.[14]

A second attribute of a written culture is a particular attitude toward authorship. Throughout the Middle Ages in Europe, an author was essentially an intermediary for conveying divine inspiration or a commentator on the writings of earlier thinkers. Respect for authors was typically delayed until after they were dead.[15] To the extent living authors supported themselves from their writings, the money came almost exclusively through patronage. Modern authorship emerged from seventeenth-, eighteenth-, and early nineteenth-century confrontations over copyright—literally, who owned the author's original manuscript ("copy") and thus had the right to profit financially by replicating it.[16]

The newly enfranchised authors who surfaced in the early nineteenth century were now the undisputed owners of their intellectual property, that is, the expression of their ideas. (The ideas themselves remained in the public domain.) Authors had the right to be paid by those who published and disseminated their writings, along with the right of propriety, protecting their texts against manipulation or degradation by others. With these new rights came added responsibilities. Authors needed to have something original to say, or at least an original way of saying it. Authors were legally answerable for the veracity of their works and bore increased accountability for the mechanics of their finished texts.

A further component of written culture is the difference between speech and writing. Writing develops its own conventions of vocabulary, grammar, and punctuation. What's more, writing mechanics, including spelling, come to matter. As Philip Dormer (better known as Lord Chesterfield) warned his son in 1750,

> orthography, in the true sense of the word, is so absolutely necessary
> for a man of letters, or a gentleman, that one false spelling may fix a
> ridicule upon him for the rest of his life.[17]

Socially, the value of writing emerges through such expression as reverence for tangible written volumes: Leather-bound sets of the complete works of Shakespeare are more highly prized than cheap paperbacks. Writing also provides a context for social affinity. Gatherings run the gamut from book discussion groups to literary speed dating at your local bookstore.

Finally, written culture has a cognitive dimension. To read is not simply to happen upon information but to encounter ideas or turns of phrase that affect us intellectually or emotionally. Zipping through the pages of *USA Today* is a fundamentally different experience from grappling with James Joyce's *Ulysses*.

• • • CHALLENGES TO WRITTEN CULTURE: THE BIG PICTURE

How have the uses and attributes of written culture stood up over time, especially in light of new computer-driven technology?

Challenging the Reasons for Writing

In professional life, the written word still holds sway, but the medium through which documents are prepared and disseminated is being transformed. Does the change in medium alter the impact of writing?

For many years, my university issued official announcements of lectures, road closings, and such via a daily voice-mail message. Items of more lasting significance were sent as paper memoranda to physical mailboxes. When you opened a university envelope, you might pause to reflect on the printed page, perhaps sharing your thoughts with colleagues in the hall. Several years ago, communication was shifted to a single daily email, with headlines followed by quick summaries and a link for more information. A former dean's death is now sandwiched between tonight's basketball game and tomorrow's lecture on bulimia. If you neglect to scroll down the page, you miss the entry entirely.

In professional writing, the prospects of writing for a living are becoming increasingly worrisome as the publishing industry squeezes out "mid-list" books in favor of hoped-for blockbusters. As for hurdle jumping, young and old alike receive conflicting messages about what constitutes appropriate written language. Traditional prose is yielding to PowerPoint speak, which Ian Parker and Edward Tufte argue represents a cognitive style quite distinct from that required for a well-constructed, sustained, even elegant argument.[18]

Likewise, written language in the social arena is being reshaped by technology. Lovers used to pen letters to one another; children wrote home from summer camp. Today, phone calls, email, or IMs largely substitute, leaving nothing to tie up with red ribbons or place in tangible family albums. Try imagining Franz Kafka's "Emails to Milena" or Horace Walpole's collected

IMs. Friends share their own poetry or short stories electronically, while writers-without-portfolio morph into book reviewers on Amazon.com.

Who today is psychologically driven to publish—either in hardcopy or online? Growing numbers of people are turning to online vanity presses such as iuniverse or Xlibris, which guide would-be authors through the self-publishing process, collecting fees for their pains. And of course there are bloggers. Individuals are still writing, but the audience they serve often ends up being primarily themselves.

Challenging the Attributes of Written Culture

If the functions and forms of traditional writing are being reconfigured, so are attributes historically associated with written culture. Start with tools for access. In his book *Scrolling Forward*, David Levy (a computer scientist by training) asks whether people who object to reading text online are simply clinging to bound books out of nostalgia. Levy compares his childhood copy of Walt Whitman's *Leaves of Grass*, published by Peter Pauper Press, with a web version, finding virtues in both. In the end, though, Levy prefers the printed version. Not only had Whitman carefully designed his collection of poems to be experienced as a book, but Levy's personal childhood history includes his relationship with a particular copy of the poems. He holds no psychological bond with the digital text.

In *The Myth of the Paperless Office*, Abigail Sellen and Richard Harper compare paper and digital technologies in terms of their respective affordances (that is, the kinds of work or activities for which a medium is particularly well suited). In office settings, many people still find it easier to mark up actual paper by hand than to do textual annotations online, and simpler to glance through a sheaf of printed documents than to rummage through their online equivalents. At the same time, the affordances of digital technology make storing information or online searching for specific words more efficient than performing the same tasks with physical documents. While enthusiastic organizational specialists predict the rapid decline of hardcopy print, many producers and consumers of the written word may not be ready to relinquish a medium they value for both its aesthetic and practical qualities.

The uncertain future of writing on paper (or reading from it) is matched by the puzzle of how the written word will be disseminated. For half a millennium, publishing houses have determined which manuscripts are printed and what those published manuscripts look like. With notable exceptions such as Ben Jonson and John Milton, early modern authors were essentially excluded from the editing process, the task being left to compositors in the printing house.

Publishing houses still vet manuscripts, frequently massaging the logic, style, spelling, and punctuation of their authors. Despite their occasional grumblings, academics and popular writers alike have generally found the contributions of publishers to be beneficial. What happens when everyone with access to a computer can become a published author? Phase one of this scenario was desktop publishing. Phase two is cyberspace, where texts wait to be read online or downloaded. The vetting and editing jobs fall exclusively on the authors' shoulders.

One hotly debated topic is the impact networked computing is having on historical notions of authorship and copyright. A fundamental challenge in copyright law—notoriously at the heart of both British and American doctrine—has always been how to motivate authors to create new works (from which they can benefit financially) while at the same time making those works readily available ("open access") to promote the common good.[19]

Open access is a particularly salient issue wherever commitment to public interest is embedded in the cultural psyche. Nowhere have presumptions about the right to free access been clearer than in the grassroots computing community. Some of the earliest computer bulletin boards, such as the WELL, can be traced to the 1960s counterculture of hippies and communes. At the same time, a "gift" mentality in the computer world at large continues to motivate free distribution of computer code and other intellectual property that would normally be protected by copyright.[20]

In chapter 6, we talked about Richard Stallman's GNU project. The project created the GNU General Public License, the essence of which reads: "This program is free software; you can redistribute it and/or modify it.... This program is distributed in the hope that it will be useful." Users are forewarned: The software comes "WITHOUT ANY WARRANTY; without even the implied warranty of MERCHANTABILITY or FITNESS FOR A PARTICULAR PURPOSE."[21] Users may download and distribute copies, and even alter the software. The only caveat is that they need to indicate they have made changes. The assumption is that, as in the case of UNIX and Linux (both of which are open source), successive users will improve the functionality of the online offering. This same principle obviously underlies Wikipedia.

What GNU did for software, the Creative Commons is now accomplishing for other kinds of intellectual property available online, including scholarship, literature, music, and photography. Thanks to the pioneering efforts of Stanford law professor Lawrence Lessig and his collaborators,

Creative Commons licenses were launched in late 2002. The goal was to find middle ground between strict interpretation of copyright ("All rights reserved") and loosely defined statements of "free" software licensing.[22]

Creative Commons licenses offer writers and artists a range of options regarding the kind of rights they wish to maintain or give away. Among the choices:

- let others download and share your work, but not permit them to change it or use it commercially
- allow others not only to download and share, but also to remix or tweak, as long as you are credited and users license their new product under the same terms you have
- permit downloading, sharing, and revising (again, as long as you are credited), but also authorize those creating derivative works to profit commercially

Designers of the Creative Commons web site have taken great pains to craft clear text and imagery for users of content covered by Creative Commons copyright, identifying precisely which use of that material is legal. Realistically, though, the bulk of Internet users continue to see online content as free, with no strings attached. Fueled by the ease of copying (and the "gift" mentality underlying so much of earlier computer culture), "the availability of [digital information and networks] has bred a mindset that seems to regard copyrighted works as available for the taking without compensation."[23]

One approach to resolving this dilemma is to tweak the balance between property and propriety. While authors of trade books understandably put a premium on financial gain, most academics are less interested in large royalties than in publication for reasons of hurdle jumping (publish or perish) or in publishing for the record. Initiatives such as the Public Library of Science[24] provide free access to timely scholarship. Articles are covered by the Creative Commons Attribution License, which permits users to copy, distribute, transmit, and adapt the work, as long as they give proper attribution to the original author(s).

If authors and artists can't count on profiting by selling their works directly, perhaps we should seriously consider a scenario concocted by John Perry Barlow, who is cofounder of the Electronic Frontier Foundation—and former lyricist for the Grateful Dead. Barlow suggests that the band's business practices contain an important lesson for contemporary intellectual property holders. Unlike many musical groups, the Grateful Dead allowed fans to audiotape its concerts. The result was a lot of intellectual property circulating for free. With the group's growing popularity being fanned by circulation of those tapes, demand for tickets to live concerts soared.[25] This model has proven highly successful on social networking sites such as MySpace and YouTube, which give away free content (and try to generate a buzz) in the hope users will pay for additional access in the form of live concerts or CDs.

In a similar vein, Barlow suggests authors should look to make money not from selling static, finished works but from real-time performances— say, download Stephen King's latest thriller for free but pay to ask him questions online. I was intrigued to read in an ad, prominently displayed in the *New York Times* Book Review, that book groups could "Enter for a chance to win a phone call from Sena Jeter Naslund [author of *Abundance*, a novel about Marie Antoinette]."[26] While the concept might work well for media mavens who relish books tours, television interviews, and phone calls, how might social recluses fare? And what about dead authors whose copyrights (currently life plus seventy years in the United States) have not expired?

An even deeper challenge emerges when we consider the product that is constructed during a live concert (or Naslund's phone chat) as opposed to the creation generated through a studio recording or published book. The latter are edited; the former are not. Unlike live performances of the "same" work (which may differ from one show to the next), "finished" performances or books present just one object for review.

Finally, there are questions of supply and demand. If anyone can access a work for free (or at very low cost), does the work lose its value in the public eye? Do authors of static texts become anachronisms? And if so, what are the consequences for written culture? To contemplate these consequences, we need to think about contemporary ways in which we value—and evaluate—the written word. To do so, we switch from a macro to a micro perspective.

• • • UNDER THE MICROSCOPE: FIVE FACETS OF MODERN WRITTEN CULTURE

A montage juxtaposes images, each with its individual integrity but, taken together, offering a whole bigger than the sum of its parts. Our montage of challenges to the ways we now value and evaluate writing contains five vignettes: Text in the Fast Lane, Flooding the Scriptorium, The Print Paradox, Snippet Literacy, and Vapor Text. Taken together, they presage fundamental changes to the existing model of print culture.

Text in the Fast Lane

Why were all the students fidgeting? Recently, I showed a graduate class the first installment of a BBC television series keyed to the book *The Story of English*.[27] Produced in the mid-1980s with a leisurely narration by Robert MacNeil, the videos had proven a highlight of courses I had offered a decade earlier. MacNeil's narrative had not become less inviting over time nor the history less vividly presented. Yet in the interim, students' notions of how long it should take to tell a story had drastically shrunk.

James Gleick's book *Faster* aptly identifies the problem in its subtitle: "The Acceleration of Just About Everything."[28] We chafe at waiting for an elevator. We want questions answered right now. Deborah Tannen argues that fast talking is now ubiquitous among teenagers and on many television sitcoms.[29] And we need to write quickly to respond to those dozens of daily emails.

Has hastening the rate at which writers create text undermined the attributes of written culture, especially the cognitive depth that writing (and reading) can bestow? If so, what are the consequences? And how did we became so obsessed with time?

Life on the Clock

Life on the clock began because Christian monks in medieval Europe needed to know when to pray. Certain religious orders called for praying at set intervals each day. Given variations in seasons and latitudes (not to mention clouds, rain, or snow), the sun or the moon was not a reliable guide.

Water clocks were introduced into the West in 807 CE, when an embassy sent by Haroun al Rashid—the fifth Abbasid caliph, whose glittering Baghdad court was depicted in the *Thousand and One Nights*—presented one to Charlemagne. But this latest Holy Roman Emperor lived in the cold north, not Arabia, and water often froze in the winter, rendering such clocks useless.[30]

Gradually, mechanical clocks were developed, and by the fourteenth century public clocks were making their way into European churches, palaces, and town squares, lending authority and order to religious, administrative, and personal activities. There was no guarantee, however, that the mechanisms all told the same time. As late as 1841, London time was ten minutes ahead of Bristol time.[31]

It was the railroad that brought a halt to local variation in setting timepieces. People wanted to know when they would reach their destination, and railroad operators needed to reduce head-on collisions between trains running on the same track. The existing method—sending riders on horse-back to warn oncoming trains that another train was approaching—was inefficient and increasingly impractical as rail traffic grew.[32]

To pace trains safely required time schedules, and for time schedules to work properly, all trains had to abide by the same norms. In 1883 standard time was established by American railroads. The following year, Sir Sandford Fleming divided the world into twenty-four time zones.[33]

The Industrial Revolution, the emergence of standard time, and the growing ease of transporting people (on trains) and people's messages (first via the telegraph and then over the telephone) were to have a profound effect on individuals and social interaction. There were now time clocks—both literal and figurative—to be punched. Real-time communication over distances, which the telegraph and telephone provided, often decreased opportunity for reflective decision making.[34] In fact, an etiquette guide written in 1914 counseled against issuing dinner invitations over the phone (rather than in writing, as was traditional) because "the person invited [by phone], being suddenly held up at the point of a gun, as it were, is likely to forget some other engagement" or simply feel there is no choice but to accept.[35]

Epitomizing the new modern obsession with time was Lewis Carroll's White Rabbit in *Alice's Adventures in Wonderland*, which appeared in 1865. Rushing past Alice, the rabbit declared, "Oh dear! Oh dear! I shall be too late." He then reached into his waistcoat-pocket, extracted a large watch, and after looking at it, hurried on, troubled about being late for the Tea Party.

Time became inextricably associated with money when the American engineer Frederick W. Taylor demonstrated that by breaking the production process into distinct, timed tasks, fabrication of goods could be stream-lined.[36] Henry Ford, who hired Taylor to conduct time and motion studies, prospered, while Charlie Chaplin's 1936 movie *Modern Times* offers a poignant critique of Taylorism.

Moving Text into the Fast Lane

The virtues of saving time have historically been an important motivation for devising techniques to speed production of written text. One early strategy, practiced by medieval scribes, was to use abbreviations.[37] With fewer distinct characters to be copied, fewer animal skins were needed to produce a manuscript.

The coming of the telegraph fostered its own truncated writing style ("BROKE. SEND MONEY."), reflecting the fact that telegrams were priced by the word. The fewer words you wrote, the less you paid. In fact, businesses in the early twentieth century raised abbreviated text to a near art form, developing elaborate cryptograms for transmitting boilerplate phrases and sentences. For instance, the British Society of Motor Manufacturers and Traders created its own codebook. If the single word *ixuah* was transmitted, the recipient decoded it to mean "Quote price and earliest day of shipment," resulting in substantial savings at the telegraph office.[38]

Reducing the number of characters and words is one way to hasten production of text. An alternative is to speed up formation of the letters themselves. Cursive writing emerged in Rome by the first century CE, facilitating more rapid generation of business and private correspondence than was possible with the traditional Roman hand.[39] Centuries later, European cursive scripts that physically joined letters together accelerated the writing process, since the pen didn't need to be raised after each letter. Later, the introduction of fountain pens carrying their own ink supply stepped up the earlier writing method based on quills (which needed sharpening) and ink wells (into which the quill had to be dipped repeatedly).

The real speed revolution came about with introduction of first typewriters and then computers. Once you learned to type, you could turn out more text in a given time interval than by hand. Accurate? Not always. But fast.

Is "Speed Writing" an Oxymoron?

Except for those flying westward, there are only twenty-four hours in a day. If I take an afternoon nap, the lawn doesn't get mowed. If I'm obsessing with my Facebook page, I'm not spending that time talking with my parents. Logic suggests that if I multitask (say, working on Facebook while on speaker phone with Mom), I may shoehorn in more than my twenty-four-hour ration. An alternative strategy is to accomplish each task in less time.

A fast mowing job probably will do, but what about a speedily produced composition? Should we actually be slowing down rather than accelerating the writing process? The answer lies not simply in the physical act of producing written characters (whether by hand or machine) but in the thought processes that hopefully go into constructing sentences and paragraphs. Speed is a virtue if you're running the mile, trying to increase profits on an assembly line, or competing on a TV quiz show. But fast thinking has not historically been associated with the written word.

In the early 1960s, the classicist Eric Havelock proposed that the intellectual accomplishments of Attic Greece could be attributed, in large measure, to the development of full-fledged alphabetic writing during the eighth and seventh centuries BCE. Havelock argued that alphabetic writing enabled the Greeks to unambiguously lay out their thoughts.[40] While Havelock's claims about the primacy of the alphabet are debatable, the more basic notion that writing facilitates clear, logical, comparative, reflective thinking remains on firm ground.

In "The Consequences of Literacy," Jack Goody and Ian Watt identify the intellectual and social consequences of writing, specifically in sixth- and fifth-century Greece:

> Literate societies...cannot discard, absorb, or transmute the past [in the ways oral societies can]. Instead, their members are faced with permanently recorded versions of the past and its beliefs; and because the past is thus set apart from the present, historical enquiry becomes possible. This in turn encourages skepticism; and skepticism, not only about the legendary past, but about received ideas about the universe as a whole. From here the next step is to see how to build up and to test alternative explanations: and out of this there arose the kind of logical, specialized, and cumulative intellectual tradition of sixth-century Ionia [precursor to the Attic Greek intellectual tradition].[41]

What Goody and Watt are arguing is that the kind of philosophical inquiry we see in the likes of Plato and Aristotle—inquiry that challenged received truths; inquiry that looked for logical relationships between ideas—was made possible by the physical ability to scrutinize historical accounts and propositions in a written form, coupled with reflection upon what was recorded.

In a similar vein, the historian Elizabeth Eisenstein suggests that centuries later, printing technology encouraged readers to reflect upon (and critique) the structure of other people's arguments. Thanks to the new availability of multiple copies of texts, scholars could sit in a single library and compare the writings of diverse authors rather than needing to trek from one repository to another (the medieval custom) to view manuscripts seriatim.[42]

Where Havelock, Goody and Watt, and Eisenstein point up the impact writing may have had upon cultural practices and understanding, it was the sixteenth-century humanist Desiderius Erasmus who proposed that individuals could strengthen their minds through guided use of the written word. In his manual *On Copia of Words and Ideas*, Erasmus counseled young men to read the works of great (inevitably dead) writers and then copy out important passages into a commonplace book, following an older medieval tradition.[43] These passages were to be organized into conceptual categories, committed to memory, and then incorporated through paraphrase into the young man's own thinking and writing. The Renaissance commonplace movement, of which Erasmus was the best-known proponent, thrived up into the nineteenth century, with a gentleman's commonplace book serving both as a vehicle for and a chronicle of his intellectual development. The initial scribal act was a necessary component in this stepwise development in the life of the mind.

Copying longhand takes time. Those of us old enough to remember doing library research using small note cards rejoiced with the appearance of copy machines by the late 1960s. For the exorbitant fee of twenty-five cents a page, you could save yourself a lot of writer's cramp. As academics learned to type, first on typewriters and then on computers, we transcribed ever-larger amounts of text—when we weren't applying yellow highlighter to our mounds of articles, now printed out for a pittance. Text was back in the fast lane.

And more's the pity, says writer Nicholson Baker, who describes how he copies out passages longhand when he wants to understand or reflect upon the words of others:

> Reading is fast, but handwriting is slow—it retards thought's due process, it consumes scupperfuls of time, it pushes every competing utterance away—and that is its great virtue, in fact, over mere underlining, and even over an efficient laptop retyping of the passage: for in those secret interclausal tracts of cleared thought-space ... new quiet racemes will emerge from among the paving stones and foam greenly up in places they would never otherwise have prospered.[44]

Erasmus would have understood.

In the twenty-first century, is this measured pace any longer possible? Given how much writing we are doing these days, perhaps not. To get a handle on the problem, we turn of all places to George Bernard Shaw's *Pygmalion*, known to many through Lerner and Loewe's musical *My Fair Lady*.

Flooding the Scriptorium

Act II. Having learned that his daughter Eliza has taken a taxi to the home of Henry Higgins, Alfred Doolittle (her father) comes to collect her—or perhaps cash in on her good fortune. Doolittle proposes that Higgins may keep Eliza, in exchange for a five pound note. Higgins's friend Colonel Picking challenges Doolittle's pecuniary bid:

> PICKERING: I think you ought to know, Doolittle, that Mr. Higgins's intentions are entirely honorable.

> DOOLITTLE: Course they are, Governor. If I thought they wasn't, I'd ask fifty.

Higgins and Pickering are incensed:

> HIGGINS [*revolted*]: Do you mean to say, you callous rascal, that you would sell your daughter for fifty pounds?
>
> DOOLITTLE: Not in a general way I wouldn't; but to oblige a gentleman like you I'd do a good deal, I do assure you.
>
> PICKERING: Have you no morals, man?
>
> DOOLITTLE [*unabashed*]: Can't afford them, Governor. Neither could you if you was as poor as me.

A garbage man by trade, Doolittle was indeed poor. If you are awash in the problems of securing daily sustenance, morality may take a back seat.

Today we are awash with written language. What might we be trading off in return?

Some of my colleagues enthuse over what they are calling an epistolary renaissance. Thanks to computers, young people are generating increasing mounds of written text. The occasion might be email, online chat forums or diaries, IM or blogs, rather than the Great American Novel. But what matters (so it is said) is that the next generation is doing a lot of writing.

Really? When properly nurtured, sustained writing experience can lead to both skill and a sense of personal and intellectual empowerment. However, just as singing off-key in the shower each morning doesn't increase your chances of making it to La Scala, merely churning out text is hardly the best way to improve your writing.

Once the computer turned us all into typists, the ever-growing online and mobile options engendered yet more text. I have come to call this phenomenon "flooding the scriptorium." Given all the writing we increasingly are doing, can we any longer afford, Governor, to pay careful attention to the words and sentences we produce? The proliferation of writing, often done in a hurry, may be driving out the opportunity and motivation for creating carefully honed text. The "whatever" attitude toward the written word may be the inexorable consequence.

In principle, there is no reason we can't do some writing the old-fashioned way: multiple drafts, time between them to think, a couple of rounds of proofreading. In practice, though, word-processing programs beckon us to push "print," while email entices us to hit "send." The convenience of electronically-mediated language is that it tempts us to make a Faustian bargain of sacrificing thoughtfulness for immediacy. In the words of the Norwegian sociologist Thomas Eriksen,

> if [email] more or less entirely replaces the old-fashioned letter, the culture as a whole will end up with a deficit; it will have lost in quality whatever it has gained in quantity.[45]

Eriksen tells the story of an Internet company official who made no apologies for errors in materials the company issued. Instead, he informed a group of Scandinavian journalists that since Internet journalists had to work very fast, rarely taking time to check sources, it was now the job of the reader, not the writer, to assume this responsibility.[46]

What about traditional writers? Consider a story on the front page of the *New York Times* in early April 2007. The piece was on the impressive talents and accomplishments of high-achieving American female high school seniors. One of the students was described as "a standout in Advanced Placement Latin and honors philosophy/literature who can expound on the beauty of the subjunctive tense in Catullus and on Kierkegaard's existential choices."[47]

"Subjunctive tense?" The term *tense* refers to time, as in "past, present, or future." The term *mood*, which is what the author of the article needed, refers to the attitude or perspective of the grammatical subject regarding the rest of the sentence.[48]

Perhaps the leader of the Internet company can be dismissed as arrogant, but I puzzled over the faux pas in the *New York Times*, a newspaper rightly respected for its editorial integrity. I don't hold the reporter responsible for knowing off the top of her head the difference between tense and mood. However, between deadlines (text in the fast lane) and a flooded scriptorium (the story ran to thirteen pages when printed from online), it seems that neither the reporter nor her editor paused to ask, "Do we know what we are talking about?"

Then there's the case of the late Stephen E. Ambrose, director of the Eisenhower Center for American Studies, founder of the D-Day Museum, and author of twenty-four books, many of them bestsellers. Respected for his thorough research and incisive analyses, he was widely praised as a media consultant and historical expert. In January 2002, Ambrose was accused of plagiarism, a charge he acknowledged as valid. Why would a capable scholar stoop to plagiarizing? The reason, said Ambrose, was time—he was in too much of a hurry to check whether the notes he used were made in his own words or directly lifted from published works written by others.[49] Two hundred years of written culture requiring authors to have something original to say went down the tubes in the name of pressure to flood the scriptorium.

In the interest of full disclosure, I admit to my own vulnerabilities when facing the slippery slope of too much writing and not enough time. Were you a fly on the wall of my study, you'd hear me grumble in resentment at having to write so much email and then proofread it all. Sometimes I feel I'm squandering my energies on producing ephemera, reducing the time (and concentration) available for serious writing. Given a choice, I always prefer a phone conversation or leaving voicemail rather than writing an email.

There's more. I confess to sometimes neglecting to capitalize letters when doing Google searches and to relying on spell-check to tell me if an expression (such as "spell-check") is two words, hyphenated, or one word. I'm increasingly prone to typing out a word on the keyboard and waiting to see if spell-check repairs my errors, obviating the need to type carefully, look up words in a dictionary, or simply think.

King Lear: The Print-Out

When I was in college, my humanities professor encouraged students to purchase two copies of the classical works we were reading: one paperback to mark up and a hardback for our personal libraries. While few took him up on getting that second copy at this point in our young careers, we understood his message that the bound book is something to be treasured.

Fast forward to the present. Yes, there are antiquarians who collect rare books, and leather-bound sets of Dickens still find an audience. But the status of a personal library has changed. In the nineteenth century, owning an extensive library had become a status symbol. In fact, the nouveau riche, many of whom cared little for reading or education, often purchased books by the yard to lend their residences the air of respectability. Over the twentieth century, the number of bachelor's degrees awarded annually in the United States increased over sixtyfold.[50] You would think that reverence for reading would climb accordingly. However, try selling a house in America that has yards of built-in bookcases. Likely as not, the realtor will temporize about how easy it is to tear out the shelving and install a family entertainment center. Books are fine, but how many do you really need?

Texts are increasingly seen as fungible. The proliferation of personal computers has led to an increase, not decrease, in use of paper. We print out email, online recipes or health advice, articles from the newspaper, chapters of books—the fruits of a flooded scriptorium. Many university courses post readings online, which students often print out. Given that Nietzsche, Shakespeare, and the Old Testament are all in the public domain, why ask students to buy printed copies, when they can run off their own for the cost of paper and ink? When the assignment has been turned in or the

"Holy cow! What kind of crazy people used to live here anyway?"

examination taken, out the pages go into the trash, along with the empty Coke cans and pizza boxes.

In what sense do you own a copy of *King Lear* if the pages are in printout? Anyone who has photocopied an entire book, perhaps spiral binding it to keep the pages together, knows it is far more unwieldy to handle than a traditional bound volume. The problem only worsens with sheaves of printed articles.

What if we forgo the printing step altogether and simply read everything online, whether as an ebook or a file downloaded to our desktop? While the environmentalist in us applauds saving trees, there are cognitive and cultural trade-offs. My humanities professor had striven to nurture within his students an understanding that many written works were worth keeping: to annotate, to contemplate, and to re-read. If printouts discourage annotation, contemplation, and re-reading, online alternatives don't even leave the starting gate.

In a written culture, books in particular and reading more generally are valued as sources of learning, reflection, cultural transmission, and shared experience. Are books and reading serving these functions today?

The Print Paradox

For the year 2005, there were $2.6 billion in sales of hardcover juvenile books, up 60 percent since 2002.[51] Harry Potter obviously accounted for a sizable chunk of those revenues, but $2.6 billion (and just for hardcover) is hardly small change.

Children's books are not the only ones selling. The Book Industry Study Group reported that in 2005, U.S. publishers' net revenues totaled $34.59 billion.[52] In the same year, American publishers issued 172,000 new titles and editions. That number is nearly three times as big as for 1995.[53] In the UK, there were 206,000 new titles for 2005—a 28 percent increase over 2004.[54]

Are people reading all these books? The late Hugh Amory, senior rare book cataloguer at Harvard's Houghton Library, once mused that "perhaps the majority of the books ever printed have rarely been read."[55] I admit to having many books on my shelf that I purchased with great anticipation though still haven't gone through. What about all those hundreds of millions of other books that are printed and for which people are paying good money?

Given continuing growth in the number of books out there for young and old alike, you might expect two corollaries to follow: that the amount people are reading is also increasing, and that, consequently, people are becoming more skilled as readers.

In the United States at least, statistics tell a different tale.

Good Grades? Yes. Reading Skills? No.

For more than thirty years, the National Center for Educational Statistics (part of the U.S. Department of Education) has issued *The Nation's Report Card*, which assesses the academic achievements of elementary and secondary school students in the United States.[56] Subjects covered include science, mathematics, writing, and reading. Given all the federal attention being directed to education, you would think the scores would be rising.

Wrong.

A report on American twelfth-grade reading scores for 2005 was issued on February 22, 2007. The assessment was based on three types of activities: reading to perform a task, reading for information, and reading for literary experience. For each subtest, scores decreased between 1992 and 2005, with the biggest drop (12 percent) coming in reading for literary experience.[57]

Declines in reading scores go hand in glove with falling results for writing proficiency. According to the National Assessment of Educational Progress (NAEP), only 24 percent of twelfth graders were "capable of composing organized, coherent prose in clear language with correct spelling and grammar."[58] Before blaming students' email and IM habits, we need to look at the writing instruction those students were receiving. The NAEP found that teachers themselves often lacked writing skills and that writing assignments were commonly abbreviated, even in English classes:

> nearly all elementary school students (97 percent) spend three hours a week or less on writing, about 15 percent of the time they spend watching television. Only half of high school seniors (49 percent) receive written assignments of three pages or more for English, and then only once or twice a month.[59]

No wonder children lack better writing skills.

Children aren't doing much reading either. A study by the Kaiser Family Foundation, released in 2005, compared the amount of time children between the ages of eight and eighteen spend in various activities over the course of a typical day. Watching TV: three hours, four minutes. Reading: forty-three minutes—most of which probably involved homework.[60]

Ironically, on February 22, 2007—the same day it noted declines in reading levels—the National Center for Educational Statistics issued a second report: "America's High School Graduates." The document presented findings on graduation rates, grade point averages, and number of advanced and challenging courses (such as Advanced Placement and second-level science offerings) students were taking. Those graduating in 2005 had grade point averages about a third of a letter-grade higher than those graduating in 1990. Ten percent of 2005 graduates completed a rigorous curriculum, up from 5 percent in 1990.[61]

More advanced level courses and higher grades for American twelfth graders, but lower reading scores overall. Books galore for children, but not much interest in reading them. Yes, children are watching television, but they watched a lot of TV in 1992 as well. Perhaps we need to look at the message their parents might be sending about the value of the written word.

Who Guards the Guardians? Adult Reading Trends

On Saturday afternoons, it's impossible to find a seat in the café of my local bookstore. Customers fill the space, spilling over into the aisles and making themselves at home in spare nooks and crannies, often ensconced with piles of potential book purchases. Judging from the crowd, Americans would seem to be reading up a storm.

Again, wrong.

Start with the question of quantity. A study by the National Endowment for the Arts reported that in 2002, only about 47 percent of the respondents had read *any* fiction, poetry, or plays in their leisure time during the preceding twelve months—down from 54 percent in 1992. (Rates for reading any book—fiction or nonfiction—also declined, from 61 percent in 1992 to 57 percent in 2002.) Not surprisingly, rates of reading literature increased with education and income, and females read more than males. However, when the data are broken down by age cohort, a troubling pattern emerges. The highest reading rate (52 percent) was for those aged 45–54. Outside of those 75 and older, the lowest rate (43 percent) was among adults aged 18–24. [62]

Are younger readers simply preoccupied with other activities such as television, movies, and the Internet, and later destined to mature into active readers? A recent study of the reading skills of college-educated adults seems to squelch the maturation hypothesis. Comparing literacy levels over time for Americans who had gone through college, scores declined significantly between 1992 and 2003. In fact, in 2003, only a quarter of college graduates were deemed to have "proficient" literacy skills. (The percent for those with some graduate school training was only slightly higher: 31 percent.)[63]

If parents aren't reading (or are not understanding what they read), we may have part of the answer behind declining literacy achievement levels in their progeny: lack of role models. However, another important piece of the puzzle may lie in the sorts of reading we ask young people to do.

Snippet Literacy

A few years back, one of the items I assigned my undergraduates to read was Robert Putnam's book *Bowling Alone*. The class was discussing the effects of the Internet on social interaction, and Putnam's carefully documented analysis of social capital in America offered a good frame of reference.

The students balked. Was I aware that the book was 541 pages long? Didn't I know that Putnam had written a précis of his argument a couple of years earlier, which they easily found on the web? One memorable freshman sagely professed that people should not be reading entire volumes these days anyway. She had learned in high school that book authors (presumably fiction excepted) pad their core ideas to make money or enhance their resumes. Anything worth writing could be expressed (I was informed) in an article of twenty or thirty pages, tops.

Back in the day, assigning a book a week in university humanities and social science courses was typical (and still is in some select schools). Now, though, many of us in academia feel lucky if students are willing to sign on for our pared-down curricular Book of the Month Club. In the words of Katherine Hayles, professor of literature at UCLA, "I can't get my students to read whole books anymore."[64]

Students in the new millennium have grown up on SparkNotes, which outline the highlights of everything from *Great Expectations* to *Harry Potter and the Sorcerer's Stone*.[65] But it's not just college students cramming for exams on books they haven't read who buy into these quick alternatives. Some members of book clubs or people who want to appear "in the know" trade in old-fashioned page-by-page reading for short study guides. In the words of Justin Kestler, editorial director for SparkNotes, "Nobody's going to read that 500-page John Adams book [by David McCullough], but people still want to know what they missed and what they should retain"[66]—an exaggeration perhaps, but indicative of a troublesome trend.

To be fair, my own era had Cliff Notes, not to mention Readers Digest Condensed Books. We also relied on introductions and secondary sources when we were too busy, lazy, or confused to work through primary texts. Yet today's college crowd has available a tool we did not: the search engine.

Search engines are a blessing. Unquestionably, they save all of us vast amounts of time, not to mention their democratizing effect for users without access to substantial book collections. But there's a hitch. Much as automobiles discourage walking, with undeniable consequences for our health and girth, textual snippets-on-demand threaten our need for the larger works from which they are extracted. Why read *Bowling Alone*—or even the shorter article—when you can airlift a page that contains some key words?

Not all blame lies with our students. As high schools and colleges cajole their faculties into making greater use of the technology in which administrations have so heavily invested, professors increasingly assign series of articles and book chapters that can be made available to students electronically. Given copyright laws, we can't usually put entire books on electronic library reserve, but selections from books and journals are fair game. In the process we "helpfully" guide students to the heart of the matter we are discussing in class.

Admittedly, in the pre-online era when research necessitated opening dozens of books in hope of finding useful information, one rarely read each volume cover to cover. It's also fair to say that given how scattershot our searches sometimes were (and the inadequacy of many back-of-the-book indexes), we often missed what we were looking for. But that said, we also happened upon issues that proved more interesting than our original queries. Today's snippet literacy efficiently keeps us on the straight and narrow path, with little opportunity for fortuitous side trips.

Consider the "Find" function, which lets us search for a particular word or phrase in a document or on a web page. Alas, it seems my students often use it in lieu of reading online assignments. When I offer them links to web sites or to journal articles for which my library has a paid online subscription, they happily contribute to class discussion or post comments on our online Discussion Board. However, when there is an article or book chapter that must be scanned before going onto electronic reserves, they balk. Am I unaware that you can't use the "Find" function on a scanned document? Could I please get them a "real" online version instead?

A related issue is the precise content of text themselves. Historically, one function of books was to offer diverse members of a community the opportunity for shared experience. Part of that experience came through having access to the same texts. Once the world of print settled down to producing exact copies of texts (essentially by the early eighteenth century), you might be sitting in Boston, England, and I in Boston, Massachusetts, and we could literally be reading off identical pages. What's more, we valued an understanding of the ways a text might evolve over time. We wanted to know how Shakespeare's quartos differed from the first folio or how Whitman reworked *Leaves of Grass*. Libraries were interested in preserving earlier versions of literary manuscripts, documenting the development of a novel or short story from one draft to the next.

In recent times, however, technology has fostered a more ephemeral approach to the written word. I call the phenomenon "vapor text."

Vapor Text

Heraclitus said you can't step into the same river twice. Literally, of course, he was right. The water a holy man bathed in yesterday when he entered the Ganges is no longer the same water he bathes in today. Yet we all agree there is some persistent notion of "Ganges."

Written language can also be seen in terms of flux and permanence. Gerald Bruns speaks of the "enclosure" of print, reifying an author's words, which came about with the transition from the medieval manuscript tradition to the rise of modern print culture.[67] While earlier readers knew to expect minor differences between manuscript copies of the "same" text (due to scribal error, attempts at correcting the textual model, or introduction of the scribe's own perspective), the emergence of written culture ushered in a growing assumption that copies of the "same" printed text were, indeed, the same, down to the last capital letter or comma.

Increasingly, that assumption is now being challenged. A case in point is the textbook I commonly use in teaching a course on the principles of linguistics. In a recent edition, I had found several errors, which I took pains to point out to my students in subsequent semesters. They looked at me blankly. In their newer print run—though of the same edition, with the same publication date—the errors had been corrected. You can't step into the same text twice.

The issue of textual permanence becomes magnified with the introduction of language online. Newspapers used to publish a morning edition and a final edition. With the Internet, the notion of an edition becomes obsolete, since text can be updated continually.

Remember the AP Latin student described in the *New York Times* as being able to "expound on the beauty of the subjunctive tense in Catullus"? My son, who several years earlier had slogged through these same high school rigors, had called me the morning the article appeared, filled with righteous indignation when he spied the inaccurate grammatical term. Excitedly, he suggested, "I'll bet you can use this example for your book." I agreed, and, as is my habit, printed out the story from online, just a few hours after the hardcopy paper had arrived on my doorstep. But the offending "subjunctive tense" had already been corrected to "subjunctive mood." The grammarian in me was relieved, but I was left to puzzle over which version of the article counted as authoritative. Medieval scribes would have felt right at home.

Newspapers have long printed corrections to errors in the previous day's paper, just as earlier, books used to include errata sheets (noting errors discovered too late to be corrected on old-fashioned linotype plates). These emendations have nearly always concerned matters of mechanical editing ("Columbus discovered America in 1492, not 1942") or objective information ("The person identified in the photograph as Al Capone was Al Franken"). Confusion of "tense" with "mood" bespeaks ignorance of grammatical categories of analysis—and of journalists' responsibility to ascertain what they don't know before instructing others in print. This kind of error would have been harder to imagine in the *New York Times* decades ago, before emergence of the "whatever" attitude toward the written word.

In the mid-2000s, Wikipedia became the poster child for online vapor text. Because anyone (with a few caveats) is free to edit any page, you never know if you are stepping into the same Wikipedia entry twice. A beneath-the-surface history of such changes is available, but most users of Wikipedia are unaware these backstage versions exist.

A further example of textual fluidity involves page numbers. The norms of twentieth-century scholarship required writers to be quite precise when they referenced works written by others, including specific page numbers, especially when using a direct quotation. Language online has rendered these norms of scholarship problematic. Consider newspaper articles that are accessed online rather than in hardcopy. The pages you might print out have no relationship to the section (and page) in the original. Sometimes the online newspaper informs you what page (in hardcopy) the piece began on, but if it's a long article, who knows what hardcopy page your quotation appeared on? Many online journals offer a similar challenge. You can always unearth the original page span, but many publishers reformat their hardcopy journals for online viewing, making exact page numbers difficult if not impossible to procure.

Should precise page references matter? Yes, if your norms of objective research include the ability for another person to pinpoint your textual findings, much as scientific experiments must be replicable for us to countenance their results. If we adopt a "whatever" attitude toward the mechanics of written texts, a similar nonchalance regarding citations is hardly surprising. In fact, it has becoming fashionable among some postmodern literary critics (and teachers of rhetoric) merely to daisy-chain the names of several authors ("Marx, Foucault, Habermas") whose writings are presumably relevant to the topic at hand, without making reference to specific works.

The notion of explicit reference is being replaced by general allusion. Yet woe be unto today's authors who offhandedly throw out names such as Ludwig Wittgenstein (in the philosophy of language) or Leonard Bloomfield (in linguistics), both of whom radically altered their intellectual positions over the course of their careers. For Wittgenstein, compare the approaches articulated in the *Tractatus* and in *Philosophical Investigations*. In Bloomfield's case, his 1914 book on language was heavily influenced by German mentalist models of psychology, while the more famous 1933 *Language* affirmed a behaviorist orientation.

What if we take the notion of vapor text yet a step further? What if we do away with text altogether?

• • • THE FUTURE OF WRITTEN CULTURE

Remember the phonograph? Today you would be hard pressed to find stores selling needles for playing records on a stereo, now that the technology has been put out to pasture by compact discs and iTunes. But when the phonograph was new, some people envisioned it might supplant the written word.

Thomas Edison's invention in 1877 was designed as a recording device into which businessmen could dictate letters without the aid of a stenographer. The resulting etched cylinders would then be mailed to the intended recipients, creating a written record that bested an ephemeral telephone conversation. A decade after Edison's first scratchy recording of "Mary Had a Little Lamb" on a sheet of tin foil, the machine was refined enough to render human speech accurately.[68]

By the late 1880s, the era of the music recording industry was still a few years in the offing. Not until 1890 were coin-operated cylinder phonographs installed in saloons—the precursor of the juke box and, eventually, the home record player. Yet already, the air was charged with possibility. Writing in the *Atlantic Monthly* in 1889, an acquaintance of Edison named Philip Hubert enthused about the potential of the phonograph: "As a saving in the time given up to writing, the phonograph promises to far outstrip the typewriter" and "I really see no reason why the newspaper of the future cannot come to the subscriber in the shape of a phonogram." In fact, Hubert mused that

> It is even possible to imagine that many books and stories may not
> see the light of print at all; they will go into the hands of their readers,
> or hearers rather, as phonograms.[69]

In that same year, another vision of the phonograph's future appeared, this time written by Edward Bellamy. In his 1888 utopian novel *Looking Backward*, Bellamy had explored what a socialist America might look like. Now he envisioned the future of print. His story, published in *Harper's New Monthly Magazine*, was "With the Eyes Shut."

The tale begins with a man boarding a train to visit a friend. The protagonist finds himself (first on the train, then when he reaches his destination) in a world in which there are essentially no books. Instead, the notes we leave for our spouses, the novels we read to pass the time, our morning newspaper have all been replaced by devices that resemble MP3 players onto which podcasts have been downloaded. Reading has given way to listening on one's "indispensable," which people carry with them at all times, much like today's mobile phones.

The advantages of the portable phonographic devices were many: People's eyesight (and posture) improved, since they no longer strained to read text or hunched over in the process. The drudgery of writing letters was alleviated, and you could review incoming correspondence at your leisure (much like listening to voicemail). Mothers didn't need "to make themselves hoarse telling the children stories on rainy days to keep them out of mischief." Instead, children could listen over and again to stories on their indispensables (a role played today by CDs, television, and the computer). In Bellamy's imagined universe, children in school

> are still taught [reading and writing]; but as the pupils need them little after
> leaving school—or even in school, for that matter, all their text-books

are phonographic—they usually keep the acquirement [of reading and
writing] about as long as a college graduate does his Greek.

Hardly a compliment to those college graduates. Bellamy envisioned that

Students and men of research, however, will always need to understand how to read print, as much of the old literature will probably
never repay phonographing.[70]

Written culture reduced to antiquarianism.

At the end of the tale, Bellamy comforts his audience that the new world
of reading "with the eyes shut" was just a dream. At the close of the nineteenth century, literate culture was booming. The rise of mass literacy was
matched with a massive production of books.[71] Even when the phonograph
became nearly as ubiquitous as the landline telephone, Bellamy's vision of
sound replacing print did not materialize.

And today? Whatever eventually becomes of modern written culture, it
seems unlikely that its material manifestations will be disappearing any time
soon. People will still read and write, paper mills will continue to do a brisk
business, and manufacturers can count on making bookcases for years to
come. Despite the growth of open source and Creative Commons licenses,
there is no immanent threat to authorial copyright on published works that
have substantial sales potential.

And yet voices from a number of quarters foresee the importance of
fixed, printed works diminishing in favor of what people such as Ben Vershbow, a fellow at the Institute for the Future of the Book, call the "networked
book":

With each passing year, our culture moves ever further from the familiar
rhythms and hierarchies of print into a vast network of machines. . . . [W]e're
headed into a fully networked culture where words and documents
are constantly in motion and conversation is the principal mode of inquiry. We're learning to read and write all over again.[72]

In 1995 William Mitchell disparagingly described books as "tree flakes encased in dead cow."[73] The same year, Nicholas Negroponte envisioned a post-
information age in which the newspaper we receive is uniquely personalized
to our interests, rather than a document shared across readers.[74]

At least for now, Amazon.com and Barnes & Noble continue to do land-
office business in marketing the printed word, bound between covers. The
issue is what roles reading and writing, books and paper will assume in the

cultural life of the coming decades. Among the questions whose answers remain uncertain are these:

- *Reading*: How much reading will be done online versus in hardcopy? How many people will be "serious," patient readers?
- *Writing*: How much writing will be done manually (with pen and ink or at a keyboard) and how much through voice recognition devices? How many people will write how much? About what? In what style?
- *Authorship*: Will the late-eighteenth-century model of authorship be replaced by one with different assumptions about the need for in-dividual creativity? Will new forms of marketing or even patronage be necessary to support people trying to write for a living?
- *Copyright*: Will copyright be replaced by licensing or open source? Will traditional notions of copyright be applied to some works (such as trade books) but not others (for example, scientific articles)?
- *Publishing*: Will books in the future largely be published only on de-mand? Given increases both in the rate of self-publishing and in tra-ditional publishing house costs, will authors become solely responsible for editing and formatting their works?
- *Language Standards*: Are we entering an era in which the mechanics of written text are viewed as less important than we have believed them to be over the past 300 years? If so, should—or can—we reverse the "whatever" attitude?

One plausible scenario is what we might call "print culture sans print." Writing might continue to be culturally valuable, but handwritten missives or printed codices would decline in importance. Under this scenario, we would become increasingly comfortable relaxing with ebooks or studying complex texts online. We might learn to produce well-edited works without resorting to printing out physical copies to mark up by hand, and could expect developments in computer hardware and software to facilitate an-notating online text so as to rival the affordances of paper.[75]

This scenario would encourage some additional changes in our notion of written culture. Printed books that continued to be produced might become essentially collectors' items; concerns about spelling and punctuation could slacken (following the present trend) without denying the importance of writing as a cultural artifact. We can imagine a society in which many of the values of print culture would be maintained without relying primarily upon familiar print technology and editorial assumptions.

An alternative scenario would be "print sans print culture." Print might remain a physically prominent component of our cultural universe, but the multifaceted aspects of Western written culture would diminish in importance. Printed works might persist but for different ends. Think of university diplomas that are still written in Latin. The text looks impressive (and highly suitable for framing), although practically none of the recipients can decipher it.

What might the future look like? It's tempting to fall back on history, to the sixteenth and seventeenth centuries, when printing presses were starting to proliferate, but before print had helped create the Western European cultural assumptions that we have identified as print culture.

Tempting, yes, but perhaps not very useful. The early modern European citizenry, who possessed minimal literacy skills and had restricted access to reading or writing materials, has little in common with a population that is overwhelmingly literate, is awash in books, and has cheap paper and pens—and computers. The future of written culture will be a product not only of education and technology but of the individual and social choices we make about harnessing these resources.

10

• • • The People We Become

The Cost of Being Always On

You sent your mom flowers for Mother's Day and want to be sure they have arrived. Because they are a surprise, you don't want to call her and ask. The solution? Track the delivery online.

The same digital technology that enables us to track packages also makes it possible to track each other. Unless we turn them off, our mobile phones ring and the BlackBerry vibrates when we are meeting with professors, apologizing to girl friends, or sitting down for dinner at a four-star restaurant.

Going the next step, there are mobile phone services—popular in Japan, and making their debut in the United States—that let you know if someone on your mobile network is physically in the vicinity. Say you're at a Nationals baseball game in Washington. A helpful message pops up on your phone that your friend Matt is also in the stadium. But the same kinds of services inform a fourteen-year-old's mother that she's not at school working on a project, as she dutifully reported, but actually at the mall.[1]

Technology has always been Janus-faced. Automobiles are convenient personal modes of transportation, but they consume vast quantities of fossil fuels and kill more than 43,000 people a year in the United States alone.[2] Refrigeration eliminated the need to go to the market each day but also meant that the food we eat is now less fresh. Modern language technologies enable a farmer standing in the fields outside the tiny Spanish town of Peñalba de Castro (abutting the old Roman theater of Clunia) to call San Francisco so his homesick daughter can listen to the local goats being led home for the night, but the same tools have herded us into a landscape where we are increasingly available as communication targets and we incessantly strategize how to control social contact.

In this final chapter, we look at the costs of being "always on" in a networked and mobile world. These costs can be measured in personal terms, ethically and cognitively, and with respect to social interaction.

• • • PERSONAE OF THE YEAR:
CONSEQUENCES OF LIFE ALWAYS ON

I was in Chicago, riding the CTA's 55th Street bus from Midway Airport. The route, which covers an ethnically diverse part of town, is not noted for a silent ridership, so the ad for US Cellular, plastered just behind the driver, was particularly appropriate: "You'll lose 163 minutes getting hit on while riding the bus, but 0 minutes when your boyfriend calls and saves you."[3] Mobile phones were designed as communication tools, but as we saw in chapter 7, they are also handy instruments for halting unwanted conversations (with friend and stranger alike).

The Me Generation Meets the Information Age:
Personal Consequences

Mobile phones are just one technology of many that can signal "Keep Out." But contemporary communication technologies also facilitate our becoming public exhibitionists. Commenting on the personal consequences of blogs, social networking spaces, and the like—what he calls ExhibitioNet, Robert Samuelson writes that "The same impulse that inspires people to spill their

© Steve Kelley. By permission of Steve Kelley and Creators Syndicate, Inc.

guts on 'Jerry Springer' or to participate in 'reality TV' shows has now found a mass outlet."⁴ Or, in the words of journalist Michael Kinsley,

> There is something about the Web that brings out the ego monster in everybody.... [D]enizens of the Web seem to lug around...deeply questionable assumptions about how interesting they and their lives might be to others.... On the Internet, not only does everybody know that you're a dog, everybody knows what kind of dog, how old, your taste in collars, your favorite dog food recipe.⁵

Social networking sites such as Twitter and Facebook let users post up-to-the-minute chatter on what they are doing ("Just finished corn flakes for breakfast. I'm about to wash the bowl.").

Time Magazine, in naming "You" the Person of the Year for 2006, declared on its cover, "You control the Information Age. Welcome to your world."⁶ But *Time* was not harkening back to the "me" generation of the 1980s. Instead, it was heralding what we might call the Personae of the Year, whereby users of information technology not only spend their hours reading blogs and watching YouTube videos but also packaging themselves however they please to put out on the Net, reality be damned.

How does our relentless access to others—and them to us—affect us personally? One obvious result: It leaves many of us exhausted. Another is that this constant contact oftentimes makes us inefficient. Genevieve Bell relates the sentiments of a small business owner in India:

> my mobile phone makes me mobile, but less efficient. When we had just one phone, and no phone in the factory, and none in the office at all, I felt more efficient.... [Now,] if I forget something, I can just call....I spend more money, I am always available, I get nothing done.⁷

A more subtle impact of nonstop communication is that it can, paradoxically, contribute to a sense of loneliness or anomie. Many young people fill the times they are alone (walking across campus, waiting in an airport) by calling or texting friends. The goal is not to share information or even say hello but to avoid being by themselves. Thomas Eriksen speaks of an increasing number of people "fill[ing] the slow gaps by talking on mobile phones while walking down a street or waiting for a traffic light to change."⁸

Half a century ago, the sociologist David Riesman noted a change occurring in the character of Americans. Where earlier they had been "inner directed" (reflecting their adherence to the norms of adult authority), now those members of the middle and upper-middle class who populated urban areas were increasing what he called "other directed" (meaning that their behavior was more commonly shaped by their peer group). The result, said Riesman, was that the middle-class American (vintage circa 1950) "remains a lonely member of the crowd because he never comes really close to the others or to himself."[9]

Riesman's notion of the lonely crowd, of people surrounded by others yet nonetheless quintessentially isolated, has reverberated over the decades. As we saw earlier, talk radio (described as one of the few public media allowing "for spontaneous interaction between two people"[10]) has long served as an antidote for loneliness in urban and rural landscapes alike. In the age of blogs and social networking sites, rather than listening to talk radio, we savor reading or viewing highly personal profiles.

The phenomenon known as lonelygirl15 is an archetypical example. In June 2006, a waif-like creature, who said she was a sixteen-year-old named Bree, began posting a video blog to YouTube. As the series unfolded, she shared her teenage thoughts and escapades, including the story of her first kiss—which has been viewed over a million times. The videos were followed by a MySpace page. (As of May 2007, lonelygirl15 had over 17,000 MySpace "Friends.") The only problem is that lonelygirl was hardly lonely or a teenager. Rather, she turned out to be a twenty-something actress whose vignettes were designed as fodder—and media buzz—for a future movie. Even when she was outed four months later, her fans kept watching and reading.[11]

Knowing Right from Wrong—and Just Plain Knowing: Cognitive Consequences

If I pick up a chocolate bar at the local convenience store and leave without paying, we would all agree I have done something wrong. Just so, if I take one of Joan Didion's literary essays, sign my name, and turn it in for a Contemporary Criticism class, I am obviously guilty of plagiarism. But what about using the Beatles' "Strawberry Fields" as background music for a video I post on YouTube (without asking permission, of course) or pulling a political cartoon off the Internet and reproducing it on T shirts that I sell to raise money for charity (again, without asking if fees need to be paid)?

Few of us keep up with the growingly involuted relationship between fair use and violation of intellectual property law, particularly given the twin contemporary mind-sets of open source and collaborative learning (a pedagogical strategy that dominates contemporary American education). A recent incident at the Duke University business school makes an excellent case in point.

In late April 2007, it was announced that nearly 10 percent of the business students slated to graduate in 2008 had cheated on a take-home final examination. These were not grade-grubbing fraternity undergraduates but MBA candidates, with an average age of twenty-nine and an average of six years of corporate work experience. Were the students just a bad lot or might there be other factors at work?

Michelle Conlin, writing a "Commentary" piece on the incident for *Business Week*, helps contextualize the problem: This is, after all,

> a generation that came of age nabbing music off Napster and watching bootlegged Hollywood blockbusters in their dorm rooms. "What do you mean?" you can almost hear them saying. "We're not supposed to share?"[12]

You say, Napster and bootlegged movies were illegal. Yet, muses Conlin, the accused might respond that "Teaming up on a take-home exam: That's not academic fraud, it's postmodern learning, wiki style." In fact, Robert I. Sutton, a professor at the Institute of Design at Stanford, is quoted by Conlin as a staunch defender of such ingenuity:

> If you found somebody to help you write an exam, in our view that's a sign of an inventive person who gets stuff done. If you found someone to do work for free who was committed to open source, we'd say, "Wow, that was smart."

Smart or sneaky, it's clear that information communication technologies will test both the letter and the spirit of intellectual property law for some time to come.

Moving to the cognitive domain, it seems likely that multitasking will be the nine-hundred-pound gorilla challenging users of language and communication technologies. We multitask of our own volition—or because we are asked to. A study by the New York Families and Work Institute reported that 45 percent of workers in the United States felt they were expected to perform too many tasks at the same time.[13] Are there consequences?

Studies on the cognitive effects of multitasking continue to appear, and the news is sobering. The bottom line is that at least for many cognitive tasks, we simply cannot concentrate on two things at once and expect to perform each as well as if we did the tasks individually.

Now a group of neuroscientists has the pictures to prove it. A research team at the Vanderbilt Center for Integrative and Cognitive Neurosciences recently published a study in *Neuron*, in which they used an MRI scanner to record brain activity of subjects performing mental tasks under two different conditions. In both instances, subjects were asked to press a computer key corresponding to one of eight sounds they heard and to speak a vowel corresponding to one of eight graphic images. Under the first condition, the tasks were presented essentially simultaneously, while in the second, they were presented one at a time. Performance on the second task was slower when the two tasks were presented simultaneously than when presented individually. The authors conclude that "a neural network of frontal lobe areas acts as a central bottleneck of information processing that severely limits our ability to multitask."[14]

A cascade of multitasking studies continues to indicate that one of the major issues is interruption. The intrusive stimulus breaks our concentration on the initial task at hand, and performance on that task degrades.

Sometimes multitasking only slows us down a bit. At Oxford University's Institute for the Future of the Mind, two groups of subjects (18–21-year-olds and 35–39-year-olds) were asked to use a simple code for translating images into numbers. Each group was given 90 seconds to complete the task. Youth triumphed, with the 18–21-year-olds performing 10 percent better than the older group. However, when each group was interrupted (by a phone call, text message, or IM), the older cohort matched the younger group in both speed and accuracy. In the words of Martin Westwell, who is deputy director of the institute, "The older people think more slowly, but they have a faster fluid intelligence, so they are better able to block out interruptions and choose what to focus on."[15]

Even computer professionals get sidetracked by multitasking. When interrupted while performing a computer-based task, they often have trouble getting back to work. Researchers at Microsoft tracked more than 2,000 hours of employees' computer activity to see what users had been up to. Their results established that following an interruption (such as an email or IM alert), users were distracted by the interruption itself, but also sometimes wandered off on tangents, such as responding to other email or browsing news or sports sites on the web. Subjects spent an average of nearly 10 minutes on the interruption and then another 10–15 minutes before returning to the originally disrupted task. In 27 percent of task interruptions, users took more than two hours to resume the original task.[16] Hardly a model of organizational efficiency.

Multitasking is the term most often applied to attempting to accomplish several tasks at once (or immediately seriatim). Linda Stone, however, suggests we shift our focus to what she calls "continuous partial attention." Founder of the Virtual Worlds Group at Microsoft Research, Stone coined the term in 1998 to refer to "post-multitasking" behavior, motivated by what she describes as our desire to be a live node on the communication network, enabling us to scan what and who might be out there and to optimize our opportunities.[17]

Continuous partial attention has been described as a practice turning our world "into a never-ending cocktail party where you're always looking over your virtual shoulder for a better conversation partner."[18] Others go so far as to suggest a strong kinship between the effects of multiple pulls on our attention (thanks to information communication technologies) and attention deficit disorder. ADD specialist Edward Hallowell, author of *CrazyBusy*, suggests that hopping from one task to the next gives workers some of the symptoms of ADD.[19] Neuroscientist Susan Greenfield considers the possibility that the proliferation of computers might help explain why ADD in children is on the rise.[20]

In 2004 a team of researchers from the University of Washington reported a correlation between television viewing by very young children and subsequent attention-deficit/hyperactivity disorder.[21] More recently, John Ratey, a professor of psychiatry at Harvard, has begun speaking of "acquired deficit disorder" to describe "the condition of people who are accustomed to a constant stream of digital stimulation and feel bored in the absence of it."[22] Ratey is not arguing that information communication technologies such as mobile phones and the BlackBerry are evil, but rather that they should be used in moderation. Drawing an analogy between food addiction and addiction to communication devices, Ratey notes that "food is essential for life, but problematic in excessive doses. And that's what makes breaking technology addiction so difficult."

"I'm Julie": Social Consequences

I needed to make a train reservation on the Acela between Washington and New York. Because I also had a question about refund policies, I decided to call Amtrak's 800 number rather than navigate the web site. "Hi. I'm Julie," the voice said. But of course, "she" wasn't Julie—or Jules, or anyone else. "She" was a voice recognition system. Typically, it takes two or three clearly articulated pleas of "AGENT!" to be put in queue for a human.

Social interaction is increasingly becoming virtual. Customer service on the web often means a "live chat" with someone whose voice you never hear. Libraries (including my own) helpfully offer IM reference services. The Embassy of Sweden recently launched a Second Life site where visitors navigate islands conveying the feel of the Stockholm Archipelago and move through a building based on the brick-and-mortar Swedish Embassy in Washington.

Even when we are face-to-face, our would-be interlocutors are increasingly moving in different social orbits. We all have stories of bank tellers or convenience-store cashiers who don't miss a beat in their mobile phone conversations while they take our money and churn out receipts. Thomas Friedman relates the together-but-apart story of a taxi ride from Charles de Gaulle airport into Paris:

> The driver and I [were] together for an hour, and between the two of us we had been doing six different things. He was driving, talking on his phone and watching a video. I was riding, working on my laptop and listening to my iPod. There was only one thing we never did: Talk to each other.[23]

Recall our discussion of Facebook in chapter 5, where we observed college students happily posting online birthday greetings to their Friends but not willing to be bothered with making an actual phone call. Or the college graduate who described Facebook as "a way of maintaining a friendship without having to make any effort whatsoever."[24] Humorist Dave Barry cleverly depicts the social conundrum:

> I've been known to e-mail people who are literally standing next to me, which I know sounds crazy, because at that distance I could easily call them on my cellphone. But I prefer e-mail, because it's such an effective way of getting information to somebody without running the risk of becoming involved in human conversation.[25]

What are the social consequences of always being able to pull the strings on the ways we interact with other people? When online communication was still in its relative infancy, a number of social scientists began worrying whether the new technologies would reduce the amount of time we spend face-to-face, would make us lonelier people, or both. As early as 1994, M. Lynne Markus reported that subjects in her study "were found to select email deliberately when they wished to avoid unwanted social interactions."[26] In the late 1990s, Robert Kraut and his colleagues at Carnegie Mellon suggested that the Internet led to feelings of depression and loneliness. [27] Stanford's Norman Nie reported that while use of the Internet at work may not be affecting social interaction, "Internet use at home has a strong negative impact on time spent with friends and family as well as time spent on social activities."[28]

Over the past decade, debate has continued to rage over whether the Internet (along with complementary mobile technologies) will undermine our social fabric. Barry Wellman and his colleagues at the University of

Toronto have done a number of studies indicating that the Internet supplements rather than displaces traditional face-to-face interaction.[29] Addressing Kraut's early findings that use of the Internet makes us lonely, a
researcher in Switzerland countered that

> the Internet has no negative effects on people's social networks. . . .
> [R]espondents who have used the Internet for a longer time period have
> the same number of close friends and spent as much time socializing with
> them as people who have used the Internet only for a short period.[30]

Kraut has modified his stance somewhat over time, suggesting participants in a
newer study "generally experienced positive effects of using the Internet on
communication, social involvement, and well-being," though he and his colleagues caution that "using the Internet predicted better outcomes for extraverts and those with more social support but worse outcomes for introverts
and those with less support."[31]

Canada recently released its 2005 General Social Survey, which measured
how respondents spent their time during a twenty-four-hour period. Results
indicated that people who logged more than one hour a day on the Internet
(excluding work or education-related activity) spent less time with their significant other, friends, or children than lighter Internet users. Heavy users
also devoted less time to domestic work (such as cleaning or childcare).
Notably, heavy users spent the most time alone, averaging about two hours
more a day by themselves than nonusers. [32]

Does the Internet, then, decrease the strength of our close social relationships? We need to be careful here. The Canadian report points out that heavy
Internet users were also the largest users of the telephone. What's more, online
communication through email and chat are forms of social interaction. In an
interview following release of the study, Barry Wellman stressed the importance of seeing Internet use in context. While acknowledging that the Internet
may sometimes cut into face-time with family and friends, he urged us to
recognize that the Internet is here to stay: "We're all becoming heavy Internet
users over time. . . . We lead somewhat different lifestyles now that include the
computer."[33]

One measure sociologists use to assess social interaction is the number of
"strong ties" and "weak ties" a person has. Strong ties are those core relationships with people you confide in, people you can count on to help in time
of need, people with whom you enjoy relaxing. Weak ties are friendly but
more at arm's length—people you know professionally or socially, from
whom you might seek information or with whom you might spend some
time but not part of your core circle. The question with technology is

whether the new media are affecting our relative proportions of strong and weak ties. The concern has been that if we are spending more time in virtual rather than in face-to-face communication, our weak ties may grow but strong ties shrink.

In 1985 a General Social Survey was conducted in the United States to assess what kinds of personal confidants Americans had, and how many of them. In a follow-up study almost twenty years later, a team of researchers replicated the questions relating to strong social ties.[34] The results were startling. In two decades, the number of people reporting they had no one with whom to talk over important issues almost tripled. Overall, the average number of confidants dropped from nearly three to barely two. Americans would seem to be growing increasingly isolated, an observation Robert Putnam made repeatedly in his book *Bowling Alone*, published in 2000, and a point David Riesman would have recognized from his vantage point fifty years earlier.

Why does the number of strong social ties matter? Beyond personal enjoyment from having close relatives or friends, strong social ties may help keep you alive. Over a dozen studies have demonstrated correlations between strong social interaction and decreased mortality.[35]

Assuming that such a decline in close social ties is occurring, what's the cause? In *Bowling Alone*, Putnam explored many possible explanations for what he describes as a growth in social isolation since the 1960s. One factor he singled out was the surge in hours spent watching television. Today, many people see computers as the culprit. But are they right?

A Pew Internet & American Life Project study directly took on the issue of Internet usage and the number (and strength) of social ties.[36] Reporting on data collected in 2004 and 2005, the authors found evidence of what they called media multiplexity: "The more that people see each other in person and talk on the phone, the more they use the Internet." Comparing Internet users versus nonusers, the number of strong ties was about the same across groups. However, the number of weak ties was slightly larger among Internet users.[37]

Contemporary data generally suggest that networked computers aren't reducing our number of friendships or (depending upon which study you read) the amount of time we spend with one another. Yet many individuals continue to voice concern that online and mobile technologies heighten their feeling of living in Riesman's lonely crowd. Listen to Robert Wright, a *New York Times* guest columnist:

> Twenty years ago I rarely spoke by phone to more than five people in a day. Now I often send e-mail to dozens of people a day. I have so many friends! Um, can you remind me of their names?... Technological change

makes society more efficient and less personal. We know more people more shallowly.[38]

Maybe so, though it becomes methodologically problematic to disaggregate
the impact of technology from other forces simultaneously at work. Rather
than automatically blaming the technology, perhaps we should look at some of
the cultural assumptions underlying the ways we use—or reject—particular
technologies.

In the late 1990s, Howard Rheingold began thinking about the devices
that allow him to be "always on, always connected" and asked, "What kind of
person am I becoming as a result of all this stuff?"[39] His search for an answer
led him to Lancaster, Pennsylvania, for a series of conversations with the
Amish.

For more than a century, the Amish have struggled with the question of
whether their members should be allowed to have telephones.[40] The issue of
adopting new technologies is not as simple as outsiders might think. Today's
Amish use disposable diapers, gas barbeque grills, and even some diesel-
powered machinery. Each new contrivance must be evaluated by the Amish
bishops, with one fundamental query in mind: "Does it bring us together, or
draw us apart?" Diesel machinery is not allowed in working the fields, since
use of the technology might jeopardize the social connection of families
laboring cooperatively. Just so, having a telephone in the house is forbidden.
In the words of one Amish man whom Rheingold interviewed,

> What would that lead to? We don't want to be the kind of people who
> will interrupt a conversation at home to answer a telephone. It's not just
> how you use the technology that concerns us. We're also concerned
> about what kind of person you become when you use it.

Fast forward to the end of 2005. Writing in the *New York Times*, David
Carr began a story about his new video iPod with more than a hint of guilt:

> Last Tuesday night, I took my place in the bus queue for the commute
> home. Further up the line, I saw a neighbor—a smart, funny woman
> I would normally love to share the dismal ride with. I ducked instead,
> racing to the back of the bus because season one of the ABC mystery-
> adventure "Lost" was waiting on my iPod.[41]

Like my student who turned down the volume on his mother by blocking her
on IM, Carr turned down the volume on his neighbor—at least this time, in
favor of watching a 2.5 inch screen. Now that television on our mobile

phones is becoming a reality, we can only begin to imagine what kind of people we may become as we increasingly isolate ourselves though physically remaining within the midst of others. In the words of Michael Bugeja, author of *Interpersonal Divide*, "For many users of mobile technology, community metamorphoses into elevator music. We know it is out there but are not really paying attention."[42]

If today's communication technologies give us tools for building virtual walls, they also provide opportunities to tether ourselves to one another. I never fail to be amazed at the cell phone conversations I overhear in my university's student union, in restaurants, or even in department stores. "Dad," the college sophomore asks, "where do I go to buy a stamp?" Or "Mom, I'm debating between these two sweaters. Should I get the yellow one or the green one?" Many college students today complain about their "helicopter parents," hovering over each decision and wanting to be part of every activity. Yet paradoxically, a lot of those same students are the ones initiating calls for advice or just to say "Hi." In 2005 a psychology professor at Middlebury College discovered that undergraduates were communicating with their parents, on average, more than ten times a week. Not only did students initiate almost half of the exchanges, but boys called home as often as the girls did.[43]

As the parent of a college student, I relish those phone calls, emails, or IMs that my son initiates. Yet reflecting back on my own college days, the thought of communicating with our families more than once a week would have perplexed my circle of friends. Wasn't part of the purpose of going off to college to learn to handle life on our own? To figure out where to buy stamps and choose our own clothing? To make mistakes and learn to recover from them? To grow up?

More generally, today's parents and children play out a bewildering set of roles involving keeping in touch, seeking advice, controlling the volume on communication, and monitoring behavior. Dad says you may go to the party but have to be reachable by mobile phone. Mom feels free to call you at any time, oblivious to the fact you might be in class or, well, have a life.

Plus ça change, plus c'est la même chose. More than a century ago, the protagonist in Edward Bellamy's futuristic tale that we described in the last chapter worried about the stifling effects of too much control over our children through technology:

> When the children grow too big to be longer tied to their mother's apron-strings, they still remain, thanks to the children's indispensable [the portable phonographic device], though out of her sight, within the sound of her voice.... It is all very well for the mothers...but the lot of the orphan

must seem enviable to a boy compelled to wear about such an instrument of his own subjugation.

The protagonist's friend, Hamage, agrees "that it was hard to see how a boy was to get his growth under quite so much government."[44]

A very different consequence of life in an online and mobile world is our loss of a sense of place. In the old days, if you called me from a landline in San Francisco, I knew you were in California. Today, when I see area code 415 show up on my phone's caller ID, likely as not it's my undergraduate assistant, walking across campus in upper northwest Washington, DC, telling me he'll be a few minutes late. Like so many American college students, his telephone exchange reflects the city where he (or his parents) took out the phone contract. Amazingly, when my students leave me voicemail messages with the request to call them back, not one of them seems to have thought about the fact that doing so requires my making a long distance call (complete with access code and a bill to my academic department) if I use my office telephone.

The situation with email is even more disorienting. I receive a message from my department chair. Is she in her office? At home? Halfway around the world? Since the email designation says "american.edu," regardless of her physical location, I have no way of knowing unless I ask. Even more extreme are the Yahoo or gmail accounts, which might originate from anywhere around the world. When students write to me asking for help with their projects on electronically-mediated communication, I often haven't a clue what country they are from.

Does the ability to localize our interlocutors matter? Anonymity has its virtues, but it can be problematic for the person on the receiving end. A student calls: Is he in his dorm room or already home for vacation? An email arrives from "american.edu": Do I need to answer it right away, or is the sender probably asleep by now since she is in Australia?

In a globally connected world, perhaps we will gradually abandon the mind-set that makes us ask, "So where are you?" when we receive a message. In the meanwhile, some of us continue to feel a bit of psychological free fall. Clifford Levy describes how he managed to retain his Brooklyn telephone number when he moved to St. Petersburg, Russia, by transferring the number to an Internet phone provider. Sometimes the convenience is an antidote for homesickness, as when his daughter calls friends back in New York. But other times the 4,500 mile gap seems unbridgeable. One day a woman

with a thick New York accent [called], wanting to know if I wanted my chimney cleaned. The woman was calling from Brooklyn and assumed that

she had reached Brooklyn. It was a humdrum sales pitch, but it gave me
that sense of disorientation that you feel when you are awake jet-lagged in a
strange hotel room at 3:45 a.m. For a moment, I considered whether it
might be a good idea to get my chimney cleaned before I realized that
I no longer had a chimney.[45]

A final consequence of our 24/7 connectivity is what we might call "the
end of anticipation." In much earlier eras, before computers, telephones, or
even the telegraph, it might take months for a letter to reach its recipient. So
much might happen before we saw one another again—deaths, births, personal transformations. People saved up stories to share and invested psychological energy in anticipating the next reunion. Coming together meant not
only physically reuniting but engaging in an unfolding of what had transpired
during the period of separation.

With the introduction of telecommunications, not to mention highway
systems, high-speed trains, and air travel, we spend increasingly less time
apart from one another, either physically or virtually. Pictures of newborn
grandchildren fly across the Internet before Mom gets home from the hospital. Your daughter's Facebook photo album reveals her newly pierced
eyebrow, softening the blow (and probably attenuating the discussion) when
she arrives home for Thanksgiving. Your son called you daily from his class
trip to Italy (thanks to the GSM phone and American calling card with which
you supplied him). When he returns, there's little point in asking, "So how
was it?" You've already heard.

When we are always on, we have the ability to live in other people's
moments. Relationships can be maintained through running discourse rather
than reflective synopsis. Absence may or may not make the heart grow fonder,
and the end of anticipation could end up being nobody's loss. However, as
with the erosion of our sense of place while communicating, there is one
thing of which we can be sure: Contemporary language technologies are
poised to redefine our longstanding notions of what it means to communicate
with another person.

• • • CAUSES AND CURES

How much of the blame for personal, cognitive, and social change associated
with new language tools really can be laid at the feet of the technologies
themselves? In many instances, we need to ask which came first: the technological development or changes in social practice?

Blaming the Messenger

For more than a decade, the public has tended to accuse online and mobile technologies of undermining our language and social fabric. Indeed, communication technologies have proliferated, and both language and society are undergoing palpable transformations. Yet as David Hume taught us long ago, constant conjunction has no necessary relationship with causation. It behooves us, then, to reflect upon change in language and social patterns independent of the technology.

Earlier we discussed shifts in attitudes toward both spoken and written language that predated the personal computer revolution. John McWhorter, in *Doing Our Own Thing*, suggests that Americans have lacked reverence for their language for at least several decades. Television's threat to turf out books began back in the days when computers were still operated by punch cards and people were not yet allowed to own their own telephones.

In the United States, adolescents and young adults have long found ways to control the volume on their communication with others. Teenagers were locking themselves in their rooms to keep family members out decades before IM (on which they can block Mom) and mobile phones (on which they let her calls go to voicemail). In Japan, *keitai* (mobile phones) are often blamed for undermining family bonds. However, the amount of time that young people are spending at home has been declining over the past twenty

Don Wright. Courtesy of the artist.

years—long before the proliferation of *keitai*.[46] Japan has also been experiencing a diminution of strong social ties since the 1970s: "there has been a decrease in people of all generations who want 'comprehensive' relationships with relatives, neighbors, and co-workers, and an increase in people who want 'partial' or 'formal' relationships." Similarly, "given the choice between 'being a good friend but not too intimate' and 'being completely open and intimate with friends,' the larger the population of the area, the more people tended to answer that they preferred the former."[47]

Within the United States, a number of voices have been trying for some time to figure out why Americans have been turning away from social engagement with one another. Networked computers and mobile phones are not the primary culprits, since the phenomenon predates them. In 1991, Kenneth Gergen argued in *The Saturated Self* that a constellation of factors— only one of which was email—were driving us into "an ever-widening array of relationships" that result in the diminution of our individual sense of self. John L. Locke, writing in *The Devoicing of Society* in 1998, suggested that people are talking with one another less and less. While some of the reasons have long histories (from urbanization to the development of literacy), other causes include information overload, answering machines, and television. Email is also on Locke's list but only the latest in the litany.

While technology may not be the root cause of recent linguistic and social transformations, it clearly has encouraged these changes. It's easier to be dismissive of others when you have caller ID than when you need to pick up the receiver to find out who is on the other end of the line. Just so, most of us are sloppier in our emails than we were in typed memoranda.

By themselves, cars don't kill people on the highway. They need to be driven. But we can reduce traffic fatalities and injuries by building better cars, setting rules of the road (including speed limits and requirements for wearing seat belts), and insisting that drivers get fundamental training in how to operate moving vehicles. In the same way, modern communication technologies don't themselves change language and society; those who use the technologies do. Are there advantages—and mechanisms—for altering these usage patterns?

The Joy of Untethering

Tuesday night, 8:00 p.m., April 19, 2007. Eight million BlackBerry users across America were brought face-to-face with their greatest nightmare: The system was down. Really down and not up again until the next morning. Redolent of their coverage of a major natural disaster or presidential primary, the newspapers gave the event much play. Life without the BlackBerry morphed into crisis mode. The chief sales officer of an insurance company explained, "I quit smoking twenty-eight years ago . . . and that was easier than being without my BlackBerry." The vice president of communication for a university hospital admitted, "I have reached the point where I get phantom vibrations, even when I'm not carrying the thing. . . . That sure doesn't sound healthy, does it?"[48]

In reading about the angst of what some jokingly refer to as CrackBerry addicts, I was reminded of an incident more than thirty years earlier in Manhattan, when a 300-block area lost access to phone service, thanks to a fire in a major switching station of the New York Telephone Company. Two professors of communication, Alan Wurtzel and Colin Turner, studied the aftermath of that 1975 fire. At the time, of course, there were no personal computers or mobile phones, so landlines were the only tool for communication-at-a-distance. Wurtzel and Turner asked:

> will feelings of isolation and uneasiness surface in people abruptly deprived of phone service? Or will they value, instead, an unexpected freedom from intrusion? Will they turn to other channels of communication and increase their face-to-face encounters?[49]

Precisely the kinds of questions social scientists might have asked of the BlackBerry shutdown.

Wurtzel and Turner interviewed 190 residents of the affected area, soon after the twenty-three-day telephone blackout ended. Here is what they found:

- 48 percent of respondents characterized the telephone as "very necessary" or "absolutely necessary" in their lives

- four out of five said they missed having the phone, especially for contacting those with whom they had primary social relationships

- more than two-thirds said that not having their telephone made them either "isolated" or "uneasy." More than 70 percent said they felt "more in control" after telephone service resumed.

What happened to them socially during those twenty-three days? Did they feel a sense of freedom? The authors reported that 47 percent of the sample "agreed that 'life felt less hectic' without the telephone" and that 42 percent "concurred with the statement that they 'enjoyed the feeling of knowing that no one could intrude on me by phone.' "[50] Face-to-face communication also increased, at least for 34 percent of those interviewed.

Growing numbers of people today at least contemplate untethering themselves from being always on. Some businesses counsel employees to cut back on their volume of email or have instituted "no email Fridays."[51] Some young adults are even closing their MySpace or Facebook accounts. Gabe Henderson, a graduate student at Iowa State University, explains why:

> The superficial emptiness clouded the excitement I had once felt....It seems that we have lost, to some degree, that special depth that true friendship entails....[In quitting MySpace and Facebook] I'm not sacrificing friends, because if a picture, some basic information about their life and a web page is all my friendship has become, then there was nothing to sacrifice to begin with.[52]

Thomas Friedman tells the story of repeatedly attempting to reach a friend in Jerusalem (via the friend's mobile phone) but each time getting no answer. Finally reaching the man (via landline) at home, Friedman inquired what was wrong with his friend's mobile phone: "It was stolen a few months ago," he answered, adding that he decided not to replace it because its ringing was constantly breaking his concentration. "Since then, the first thing I do every morning is thank the thief and wish him a long life."[53]

Learning to Listen—and to Hear

We looked in chapter 6 at the role talk radio has played in providing companionship for the lonely and in enabling the more outgoing to have their say. But sometimes talk radio rises to sophisticated dialogue and debate, proving as informative as it is provocative. In the United States, National Public Radio (NPR) has taken the lead on this kind of programming.

One of the smartest talk radio hosts in the medium's history is Diane Rehm, whose "Diane Rehm Show" on NPR attracts a educated crowd from around the world. Each day she closes her program with the words, "Thanks for listening." *Listening.* I had always interpreted this word as a stand-in for "listening in" or "tuning in" to her program rather than changing the dial.

In May 2007, Rehm delivered a university commencement address, in which her topic was listening. Rehm argued that

> In this day and age of email, voicemail, office memos and text messaging, we hardly ever hear each other in real time anymore, much less listen to each other. In fact, I believe many of us may have forgotten how to listen.

Why, Rehm asked, does listening matter?

> Listening is a form of engagement, of opening the mind and the heart to another.... [B]y listening instead of talking, we may become more attentive to the whole person, and by that very engagement, we learn more about ourselves.

What Rehm is talking about is listening so that we can hear. Hear what others have to say, and, by extension, hear ourselves think. If we are continually being bombarded by communication stimuli—and barraging others, very little listening and hearing are going on.

Connecting Responsibly

Eating nuts is good for your health, but eat too many and you may grow fat. Playing the piano is a marvelous skill, but practice incessantly and you might suffer from carpal tunnel syndrome.

Modern language technologies are invaluable aids to human productivity, social connectedness, safety, and relaxation. However, we may need to learn to use them more responsibly. When Coca-Cola was first concocted by a druggist in Atlanta, the formula included cocaine (hence the name). Only later did we learn the dangers of the ingredient. Are there linguistic or social side effects of using (and overusing) computers and mobile devices?

We've seen contemporary users complain about or fall victim to the technology, particularly under the pressure always to be connected. We learned of the deleterious effects that communication distracters such as a ringing telephone can have on IQ scores. We encountered the student who wrote, "Facebook, I hate you" for being so addictive. We have read about palliative moves such as "no email Fridays." India and Japan have a tradition of leaving your outside shoes at the door when you enter the house, clearly demarcating the boundary between home and the world. Now that email, mobile phones, and the BlackBerry follow us into our homes and even on vacation, it takes more than simply removing our shoes to leave the world of connectivity behind. What is needed is an act of will.

Thomas Ericksen suggests a distinction between what he calls fast time and slow time. Obviously, there are occasions when we need to be connected, when efficiency matters, when we must be findable. But, he argues, we also should delineate time to sit in the park and people-watch or admire the flowers. Eriksen extends the dichotomy to fast thought and slow thought—each has its place. Quick thinking may help us outrun a rattlesnake, while slow thinking is better suited to solving problems in mathematical physics.

In much the same way, we can distinguish between fast writing and slow writing—and times appropriate for each. Fast writing is fine for putting together a "to do" list, dashing off an IM to a colleague, or jotting down the outline (or even first draft) of an argument. But slow writing—perhaps even handwritten, perhaps composed at a keyboard, but definitely revised and edited—must remain the gold standard for writing text that enables us to formulate and convey meaningful analysis to others and to ourselves. The problem with contemporary writing technologies is not they enable us to write quickly but that they threaten to overwhelm slow writing. The challenge is that the convenience of email, IM, and texting tempts us to sacrifice intellect and elegance for immediacy.

The Long Arm of the Law

Another solution, of course, is to call the cops. Concerned that people should not be driving while using a BlackBerry? Make such behavior against the law.[54] Frustrated when students make excessive reference to Wikipedia on term papers? Ban it.

In recent years, Americans have increasingly been fighting back against uses of technology they find to be problematic. Despite all the money that was spent installing wireless connections in academic buildings, many faculty are nixing computers in their classrooms. Instead of letting students IM one another, watch YouTube, surf the net, or even bury their faces in their computers as they furiously type notes on the lecture, the faculty hope to increase real face-to-face discussion. Some high schools are starting to insist students check their mobile phones and MP3 players at the door before walking into a class on test day, given the obvious opportunities these devices offer for cheating.[55]

Middlebury College's history faculty created quite a stir in February 2007 when it decreed that references to Wikipedia were unacceptable in work submitted to the department.[56] The professors did not deny that Wikipedia can be a helpful resource for getting started on a subject about which you know little to nothing. But students were then required to go the next step in finding traditional refereed sources.

As the history of *keitai* in Japan illustrates, societies are capable of reconfiguring technological practices if they have the will. When *keitai* first appeared in large numbers in Japan during the early 1990s, they were essentially used for talking. By the late 1990s, the government began banning voice calls in public places such as trains and buses. The long arm of the law, coupled with public shame, were highly effective in transforming outdoor use of *keitai* from talking to overwhelmingly texting instruments.

Signs of Change

Comparatively speaking, contemporary language technologies are quite new. Popular use of email is barely two decades old; text messaging, from a few years to about fifteen (depending on what part of the world); IM, barely a decade. Blogs, Wikipedia, and Facebook are downright newbies. If it took at least thirty-five years after the invention of the telephone to decide it was reasonable to issue a dinner invitation by phone, we should hardly be surprised that best practices for networked computers and mobile communication devices are far from worked out. Similarly, it may be too soon to understand the impact these tools may be having upon us over the long haul.

I'm reminded of the early days of desktop publishing, when we suddenly found ourselves in possession of a dazzling array of type fonts and sizes, rather than just tried-and-true Times New Roman in 10 or 12 point. While most of us had no background in graphic design, we nonetheless delighted in creating font frenzies that one writer aptly described as ransom notes.[57]

Over time, our exuberance calmed down, and the documents, brochures, posters, and the like that we now create tend to be fairly staid. Similar trends may already be at work with some forms of electronically-mediated language: educators offering explicit guidelines for online research practices, college students dropping out of Facebook. Michael Bugeja says "I think we're at the very beginning of [the tech generation] reaching a saturation point" with virtual social networking platforms.[58]

One factor may be that new technologies are losing their "specialness" as the young find themselves sharing tools with their parents' or even grandparents' generation. Mom knows how to IM, Dad does texting, and Great Aunt Sarah has started her own blog. A defining moment came in December 2006 when the Central Intelligence Agency put up a Facebook Group, complete with a thirty-second promotional YouTube video, to recruit employees for its National Clandestine Service.[59] Was nothing sacred? After one month, the Group boasted more than 2,100 members. Today, every major political candidate in America has a presence on Facebook and MySpace, not to mention blogs.

Like language, technology does not remain static. At the same time, just as certain components of language hang around for centuries while others come and go, we can anticipate that some—but not all—electronic language media will have staying power. Before worrying inordinately about American teenagers sending thousands of text messages a month (other than our protesting the bill), we may be better served by waiting a bit to see if the storm blows over without need for aggressive intervention. At some point, excessive use of texting and Facebook and Wikipedia and IM may simply become yesterday's news.

• • • ON BEYOND CAMELS

In the preface, I noted my temptation to call this book *Beyond Email*, with a conscious nod to Theodor Geisel's imaginative *On Beyond Zebra!* I end our journey with a nod to the Grand Vizier of Persia.

In *A History of Reading*, Alberto Manguel recounts a tale that is sure to stir the heart of any bibliophile:

> In the tenth century...the Grand Vizier of Persia, Abdul Kassem Ismael, in order not to part with his collection of 117,000 volumes when traveling, had them carried by a caravan of four hundred camels trained to walk in alphabetical order.[60]

Given the technology of his day (camels), the Grand Vizier (or his counselors) devised an ingenious solution to the problem so many of us struggle with in gaining ready access to the contents of thousands of books.

But the Grand Vizier is hardly alone in his creativity. The way we use a technology is always a joint product of the technology's affordances and of the cultural milieu in which it plays out. The Vizier had the power to commandeer all those camels. If Lewis Carroll's White Rabbit had been created in 2008 rather than in 1865, would he simply have text-messaged the Red Queen to let her know he was running late? Or are some solutions simply not workable under any circumstances? While Queen Elizabeth always carries her purse, I would be surprised if she checks for messages on her mobile while at formal events. Our point is as old as the verse in Ecclesiastes: To everything there is a season—or, in our case, to every use of language technology, there are plausible times and places.

I would have enjoyed the Vizier's library caravan, at least for the first day or two, but then I suspect the novelty would have worn off. In much the same way, computer technologies have generated an air of excitement, but their fate can hardly be predetermined. To illustrate what I mean, I close by juxtaposing two technological moves by my university library.

Like all such libraries today, mine has committed itself to investing in online materials. As the number of electronic journals and databases increases, purchases of hardcopy materials have had to give way. Much as we cherish books, faculty and students alike have applauded the move, given that resources are finite. It's not that we love books less but twenty-four-hour virtual access more.

Several years ago, my tech-savvy library introduced an automatic book checkout system that enabled users to bypass lines at the circulation desk. Since patrons knew how to operate ATM machines, purchase airline tickets online, and navigate the library's rich web site, it stood to reason that we would take to the technology like ducks to water.

Wrong. After some initial fanfare, the machine now stands largely idle. Why?

I think the answer lies in the presence of an alternative: real people working at the circulation desk. While we conduct our business, we get to say hello to another human being, grouse about the weather, or commiserate about the latest campus scandal. Yes, we went to check out our books, but much more ended up happening.

The writer Rose Moss captured this "human value-added" in her recounting of a snowstorm in Massachusetts that kept everyone at home—perhaps spending some time with the family but also handling office work via computer and phone. Fine and good, but something was missing:

> We are gregarious animals. We want to see what other people look like when we talk to them. Sometimes technology gives clues: we can sometimes tell when someone we're talking to on the phone is smiling. But in hallways we pick up nuances like eye contact, closeness or withdrawal. We hear jokes, stories and gossip. We get the changing play that holds us in conversation.[61]

Moss goes on to argue that "What will always be absent from cyberspace is space where we meet people we did not necessarily plan to meet." The same can be said of encountering unexpected volumes when we browse actual bookstore or library shelves rather than efficiently finding books online. Yes, Amazon suggests other items we might want to buy, but that is not the same experience as ensconcing yourself on the floor or in a carrel, poring over volumes that you happen upon.

On the other hand, there are times when automatic book checkout might prove incredibly convenient, and a lot of people relish telecommuting full time. If you're good at searching online, you can discover wonderful books that your bricks-and-mortar bookseller or library doesn't carry. Just as we have individual tastes in music and clothing, one size doesn't fit all when it comes to using computers and mobile phones.

Obviously there are professions that insist you take your BlackBerry on vacation, and social circles in which it's not cool to neglect your blog. But most of us have more freedom than we realize to shape our own usage of language technologies. We have substantial say over the extent to which we multitask. We can opt to make our IM or text messages look more like speech or writing, and we can present ourselves online in guises of our own choosing. We can select between talking and texting on our mobile phones, and make up our own minds about how careful to be in our written language—online or off. We can pick and choose which library resources to use electronically and when to luxuriate with real books. If we wish, we can battle our way past "Julie" on Amtrak and talk with a live human. We can decide for ourselves whether to remain "always on."

Such choices give us ultimate control over language in an online and mobile world.

• • • Notes

Chapter 1

1. Silverstone and Haddon 1996.
2. McWhorter 2003:xxv.

Chapter 2

1. http://theworldwidegourmet.com/fruits/pineapple/history.htm.
2. "interactive written discourse": Ferrara, Brunner, and Whitemore 1991; "e-mail style": Maynor 1994; "electronic language": Collot and Belmore 1996; "Netspeak": Crystal 2001.
3. Boettinger 1977:66.
4. Campbell 1998.
5. Abbate 1999:201.
6. Nielsen/Net Ratings April 23, 2007.
7. I am grateful to Rich Ling for clarifying this point.
8. One of the exceptions is Kalman et al. 2006.
9. See Ling 2007.
10. http://web.mit.edu/olh/Zephyr/Revision.html.
11. Tom Anderson, Brad Greenspan, and Chris De Wolfe (among others) played key roles in launching MySpace. Facebook was developed by Mark Zuckerberg. YouTube is the creation of Chad Hurley, Steve Chen, and Jawed Karim.
12. http://livinginternet.com/r/ri_emisari.htm.
13. http://secondlife.com.
14. Sipress 2006.
15. Newitz 2006. For a sampling of educational use of Second Life, see http://web.ics.purdue.edu/~mpepper/slbib.
16. "IBM to Build Virtual Stores in Second Life" 2007.
17. http://www.internetworldstats.com/stats7.htm.
18. "Telecommunications and Information Highways" 2006.
19. http://www.internetworldstats.com/stats7.htm.

20. Cited in Musgrove 2007.

21. Cited in http://www.techcrunch.com/2006/07/24/instant-messaging-and-trashing-google/.

22. http://technorati.com/about.

23. Cited in Stone 2007b.

24. http://www.itu.int/ITU-D/icteye/Indicators/Indicators.aspx#. Accessed June 23, 2007. Historical statistics continue to be updated on the site—hence the relevance of knowing the date they were accessed.

25. Blumberg and Luke 2007.

26. Fram 2007.

27. GSM World. Available at http://www.gsmworld.com/index.shtml.

28. Pressler 2007. Data are from the CTIA.

Chapter 3

1. Gibson 1979; Sellen and Harper 2002; Gaver 1991.

2. Westin 1967.

3. Baron 2002.

4. Morton 2000.

5. Chapter 5 describes a similar ploy using Facebook.

6. Some people post an away message when actually sitting at their computers, enabling them to screen which incoming IMs to respond to and which to ignore.

7. Katz and Aakhus 2002.

8. See chapter 7.

9. Damos 1991; Floro and Miles 2003; Ironmonger 2003; Michelson 2005; Ruuskanen 2004; sciam.com 2004.

10. Kenyon and Lyons 2007.

11. Manhart 2004; Stroop 1935.

12. Armstrong and Sopory 1997; Pool et al. 2003.

13. Rogers and Monsell 1995; Rubenstein et al. 2001.

14. L. Brooks 1968.

15. Daoussis and McKelvie 1986.

16. Rideout et al. 2005:36.

17. Rideout et al. 2005:54.

18. Hewlett-Packard 2005

19. Hembrooke and Gay 2003.

20. Crook and Barrowcliff 2001.

21. Adamczyk and Bailey 2004; Cutrell et al. 2001; Dabbish and Kraut 2004.

22. Draganski et al. 2004.

23. Maguire et al. 2000. For a layperson's version of the story, see "Taxi Drivers' Brains 'Grow' on the Job" 2000.

24. Baym et al. 2004; Lenhart et al. 2005; Shiu and Lenhart 2004.

25. Sprint 2004.

26. Hewlett-Packard 2005.

27. Baym et al. 2004.

Chapter 4

1. Boneva et al. 2006; Grinter and Paylen 2002; Issacs et al. 2002; Lenhart et al. 2001; Nardi et al. 2000; Schiano et al. 2002.

2. Jacobs 2003; Hård af Segerstad 2002.

3. For analysis of this continuum, see Tannen 1982a, 1982b; Chafe and Tannen 1987.

4. This list draws upon my own previous work (Baron 2000, 2003), along with studies by Wallace Chafe and Jane Danielewicz (1987) and David Crystal (2001).

5. I benefited greatly from work by Milena Collot and Nancy Belmore (1996), and by Simeon Yates (1996), whose detailed comparisons of one-to-many CMC conversations with large-scale corpora of spoken and written language (including those collected by Douglas Biber, the London-Lund speech corpus, and the Lancaster-Oslo/Bergen written corpus) helped shape my thinking.

6. Baron 1998.

7. Crystal 2001:47.

8. Research by Harvey Sacks and Emanuel Schegloff laid the groundwork for most contemporary conversational analysis (Sacks et al. 1974; Schegloff and Sacks 1973).

9. See Halliday 1967; Crystal 1975; Chafe 1980, 1994, 2001; Chafe and Danielewicz 1987.

10. I am grateful to Susan Herring for urging me to pursue the issue of intonation units in spoken discourse as a way of thinking about IM.

11. As with most analogies, the fit between Chafe's notion of a (spoken) intonation unit and transmission units in IM is not precise. In IM, distinct transmissions are easy to count: You can always tell when the sender hits "Enter." With speech, dividing up a conversational turn into intonation units leaves more room for ambiguity. Despite this problem, Chafe's intonation units at least give us a place to start in our analysis of IM as a spoken or written form of language.

12. For reviews of the sociolinguistic literature, see Cameron 1998; Holmes and Meyerhoff 2003.

13. Boneva and Kraut 2002; Thomson et al. 2001.

14. Kendon 1980.

15. Coates 1993.

16. Chambers 1992; Holmes 1993.

17. Cameron 1998; Coates 1993; Eckert and McConnell-Ginet 2003; Holmes 1993; Romaine 2003; Tannen 1994.

18. Holmes 1995:2.

19. Chambers 1992; Holmes 1993; James 1996; Labov 1991.

20. Biber 1988; Biber et al. 1998; Biber and Finegan 1997.

21. Work by Palander-Collin (1999) further supports these findings.

22. Mulac and Lundell 1994.

23. Koppel et al. 2002; Argamon et al. 2003.

24. *The Nation's Report Card* is compiled by the National Assessment of Educational Progress (NAEP), under the auspices of the National Center for Educational Statistics.

25. Sproull and Kiesler 1986, 1987, 1991.

26. Herring 2000.

27. Herring 2003:207.

28. Herring 2003.

29. All the student names in examples—both here and throughout the book—are pseudonyms.

30. The best-laid research plans are sometimes no match for the vagaries of student life. Several of the student experimenters faded into the woodwork during the data-collection period, and our goal of a large balanced sample had to be scaled back. The corpus wins no prizes for elegance in research design. As an exploratory study, however, the project yielded some very clear trends, many of which have been corroborated by other investigators' research—for example, Squires (2007) and Tagliamonte and Denis (2006). Note that some of our analyses were performed on the entire set of IM conversations whereas others were restricted to comparing the 9 female–female and 9 male–male conversations (together totaling 1,861 conversational transmissions).

31. Chafe and Danielewicz 1987:96.

32. Sacks et al. 1974.

33. Gloria Jacobs, personal communication; Nardi et al. 2000.

34. Jacobs 2003.

35. For statistical details on the analysis of IM conversations described in this chapter, see Baron 2004 and Baron In press.

36. For the full coding scheme see Baron In press. Nearly all of the coding system followed standard definitions of grammatical categories (such as adjective, direct object) and functions (for example, *and* is a coordinating conjunction, while *because* is a subordinating conjunction). The overall grammatical model reflects a simplified version of early transformational grammar (Chomsky 1965), combined with terms from traditional grammatical models ("independent clause," "subordinating conjunction"), plus a few minor adaptations.

37. Some contractions appeared in the corpus, though fewer than anticipated.

38. Some EMC abbreviations and acronyms appeared in the corpus, though fewer than anticipated.

39. Emoticons are typically used in lieu of (spoken) prosodic or kinesic cues available in face-to-face communication.

40. Speech is often characterized by run-on sentences whose components are chained together with coordinating conjunctions. Whole sentences in formal writing generally do not begin with conjunctions. However, Chafe observes that new intonation units in speech are sometimes composed of independent clauses.

41. Chafe 1980:14.

42. An example from Chafe (1980:15):

Intonation Unit 1:   This time I saw a statue
Intonation Unit 2:   it looked like it was in a park

43. Chafe 1980:20. Technically, *the* is a determiner, but for purposes of our example, it functions the same way as an adjective.

44. Chafe 1980:46.

45. Ling 2005.

46. Chafe and Danielewicz 1987:103.

47. However, as with EMC abbreviations and acronyms, there were relatively few emoticons in the data, and most of these were a smiley, i.e., :-).

Chapter 5

1. Goffman 1959.

2. Jacobs 2003:13.

3. I am grateful to Lauren Squires, Sara Tench, and Marshall Thompson for assistance with the study.

4. Much has been written about the role of humor in computer-mediated communication. See, for example, Baym 1995 and Danet 2001.

5. IM Profiles are also used for this purpose.

6. I have chosen to examine Facebook here (rather than MySpace) because of its university-based roots.

7. boyd 2004.

8. Our Facebook history draws upon Cassidy 2006.

9. Cassidy 2006.

10. Cassidy 2006.

11. Cassidy 2006.

12. Both of these tools are essentially news aggregators, similar to RSS feeds.

13. "Facebook Gets a Facelift," from the Facebook site (http://www.facebook.com) on September 5, 2006.

14. According to comScore Media Metrix, 12.4 million in December 2005 versus 12.9 million in March 2006.

15. http://www.insidefacebook.com.

16. By late 2006 and then 2007, a number of academic studies of Facebook had begun to appear, including Acquisti and Gross 2006; Ellison et al. 2007; Golder et al. 2006; and Vandeen Boogart 2006.

17. Ellison et al. 2007.

18. Vanden Boogart 2006.

19. Golder et al. 2006.

20. Golder et al. 2006:3.

21. Finder 2006.

22. Average response: 3.15 on a 5 point scale, where 1 = strongly disagree and 5 = strongly agree (Stutzman 2006).

23. Acquisti and Gross 2006:8.

24. Acquisti and Gross 2006:18.

25. Gross and Acquisti 2005.

26. Stone 2006.

27. The fall 2005 study was done by Tamara Brown, Dan Hart, Kathy Rizzo, and Kat Waller.

28. comScore Media Metrix.

29. *Newsweek* July 17, 2006, 12.

30. Gonzales 2006.

31. Peters 2006.

32. Ellison et al. 2007.

33. Vanden Boogart 2006:38.

Chapter 6

1. Sharpe 1985.

2. Gatrell 1994:32–33.

3. Sydney's Speakers' Corner began in 1878. About ninety years later, the Free Speech Movement at the University of California, Berkeley, made a "speakers' corner" out of Sproul Plaza. Chicago has its Washington Square Park, and other modern speakers' corners have been established in Saskatchewan and Cardiff, Singapore and Amsterdam.

4. Useful sources on the history of newspapers include Smith 1979 and Chalaby 1998.

5. Our historical account of letters to the editor draws upon Hannah Barker's work (H. Barker 1998, 2002; H. Barker and Burrows 2002), along with Black 1987; Bourne 1966; Shaw 1985; and Sherbo 1997.

6. H. Barker 1998:38, summarizing an argument by C. Brooks 1991:57.

7. Sherbo 1997:iii.

8. Dodgson 2001:81, 135.

9. Doyle 1986.

10. Shaw 1985:xiii.

11. Shaw 1985:xi.

12. Laufer 1995:39.

13. Laufer 1995:38.

14. Brewer 1996.

15. "The Talkers Magazine Heavy Hundred Class of 2005," *Talkers Magazine Online.* Available at http://www.talkers.com/heavy.html.

16. Annenberg 1996.

17. How Americans Get Their News, Gallup Poll, December 31, 2002.

18. http://www.arbitron.com/downloads/radiotoday07.pdf.

19. http://www.arbitron.com/downloads/radiotoday07.pdf. The report indicates that Americans age twelve and over averaged nineteen hours a week listening, though we probably need to distinguish between actual listening and the number of hours the radio was simply turned on.

20. Laufer 1995:118.

21. Munson 1993:47.

22. Munson 1993:1.

23. Mark Sandell, editor for the BBC's "World Have Your Say," has written that while the show is "a news programme first and foremost," it is "the listeners who dictate the agenda." Its aim is "to tap into what people are talking about and find out what's really making YOUR news, not just ours." Yes, the show has "a vibrant and dynamic team of journalists...and presenters [hosts] who have brilliant qualities,"

but one of those qualities is "knowing when to shut up." See http://news.bbc.co.uk/go/pr/fr/-/2/hi/talking_point/world_have_your_say/4386054.stm.

24. Livingstone and Lunt 1994:101.

25. Milgram 1970.

26. Livingstone and Lunt 1994:43.

27. Turow 1974:171.

28. Turow 1974:176.

29. Avery et al. 1978:16.

30. Armstrong and Rubin 1989:90.

31. Armstrong and Rubin 1989:90.

32. Scott 1996:1.

33. Laufer 1995:15.

34. Laufer 1995:32.

35. Laufer 1995:60.

36. Laufer 1995:14.

37. D. Barker 2002:119–120.

38. http://people-press.org/reports/display.php3?ReportID=248.

39. Laufer 1995:119.

40. Podcasting serves a parallel oral function.

41. Page and Tannenbaum 1996.

42. Nardi et al. 2004:230.

43. See, for example, Owyang 2006; Lasica 2001; McNeill 2003.

44. Lenhart and Fox 2006.

45. Graf 2006.

46. British Market Research Bureau April 2006, reported by Caslon Analytics (http://www.caslon.com.au/weblogprofile1.htm).

47. http://www.caslon.com.au/weblogprofile1.htm.

48. Herring, Scheidt, Bonus, and Wright 2005.

49. Nardi et al. 2004:228.

50. Herring, Kouper, Paolillo et al. 2005.

51. Herring, Scheidt et al. 2005.

52. See, for example, Serfaty 2004.

53. Paquet 2002.

54. Krishnamurthy 2002.

55. Herring, Kouper, Scheidt, and Wright 2004.

56. Herring and Paolillo 2006.

57. Lenhart and Fox 2006.

58. Nardi et al. 2004:225.

59. Nardi et al. 2004:227.

60. Bahrampour 2007.

61. Bahrampour 2007.

62. Nardi et al. 2004:227.

63. Schiano, Nardi et al. 2004:1146.

64. Fine 2006; Ralli 2005.

65. Bahrampour and Aratani 2006.

66. http://people-press.org/reports/display.php3?ReportID=248.

67. Okin 2005:250.

68. Feuer and George 2005.

69. See Baron 2000:58–69 for a summary of landmarks in the development of English copyright law.

70. http://www.copyright.gov/circs/circ1a.html.

71. For more on wikis, see Leuf and Cunningham 2001.

72. Charles Van Doren (1962) offers an eloquent response to the question of goals in "The Idea of an Encyclopedia," his commentary on *L'Encyclopédie française*.

73. Our story of the *Encyclopédie* is based on the "Introduction" to Diderot et al. 1965 and Blom 2005.

74. This discussion draws upon Kogan 1958.

75. Van Doren 1962:25.

76. McHenry 2003.

77. http://www.dwheeler.com/secure-programs/Secure-Programs-HOWTO/history.html.

78. Rheingold 1993; Turner 2006.

79. More on the snippets issue in chapter 9.

80. http://wikipedia.org/wiki/Nupedia.

81. http://en.wikipedia.org/wiki/Wikipedia.

82. http://en.wikipedia.org/wiki/Wikipedia:Five_pillars.

83. http://stats.wikimedia.org/EN/TablesWikipediansEditsGt5.htm.

84. http://stats.wikimedia.org/EN/TablesWikipediaEN.htm.

85. I am grateful to Jack Child for his insider's guide to the work of Wikipedia contributors.

86. http://en.wikipedia.org/wiki/Wikipedia:Awards.

87. http://en.wikipedia.org/wiki/Wikipedia:Contributing_to_Wikipedia.

88. Sanger 2004. In chapter 8, we return to the question of whether a new norm of "pretty good" is replacing "excellent" or "precise" in contemporary writing.

89. Sanger 2006.

90. McHenry 2004.

91. Giles 2005.

92. Encyclopedia Britannica, Inc. 2006.

93. http://citizendium.org/fundamentals.html.

94. Emigh and Herring 2005.

95. Rosenzweig 2006.

Chapter 7

Suggested Readings on Mobile Phones

Readers interested in learning more about research on mobile telephony will find a wealth of information in the books listed below. Full bibliographic details are in the comprehensive list of references at the end of this book.

Castells, M., M. Fernández-Ardèvol, J. L. Qiu, and A. Sey. 2007. *Mobile Communication and Society: A Global Perspective.*

Glotz, P., S. Bertschi, and C. Locke, eds. 2005. *Thumb Culture: The Meaning of Mobile Phones for Society.*

Goggin, G. 2006. *Cell Phone Culture: Mobile Technology in Everyday Life.*

Hamill, L., and A. Lasen, eds. 2005. *Mobile World: Past, Present, and Future.*

Harper, R., L. Palen, and A. Taylor, eds. 2005. *The Inside Text: Social, Cultural, and Design Perspectives on SMS.*

Ito, M., D. Okabe, and M. Matsuda, eds. 2005. *Personal, Portable, Pedestrian: Mobile Phones in Japanese Life.*

Katz, J.E., ed. In press. *Handbook of Mobile Communication Studies.*

Katz, J.E., and M. Aakhus, eds. 2002. *Perpetual Contact: Mobile Communication, Private Talk, Public Performance.*

Ling, R. 2004. *The Mobile Connection: The Cell Phone's Impact on Society.*

Ling, R., and P. Pedersen, eds. 2005. *Mobile Communications: Re-Negotiation of the Social Sphere.*

1. Drawn from Agar 2003:48–66.
2. GSM World.
3. GSM World.
4. The story here is based on Agar 2003:94–101; Kohiyama 2005:61–63; Matsuda 2005a:22.
5. For a history of DoCoMo, see Beck and Wade 2003.
6. See Agar 2003:19–43, 67–69.
7. Korea has also had CDMA phones.
8. Presently I am gathering cross-cultural data on mobile phone usage patterns in Sweden, Italy, Japan, and the United States.
9. Ito et al. 2005 provide an extensive study of mobile phone use in Japan. Another book-length analysis of mobile telephony in social context is Horst and Miller 2006, examining the role of mobiles in Jamaica.
10. Okada 2005:56.
11. Bell 2005:77.
12. Donner 2005, 2007.
13. Matsuda 2005a:28, 38 n. 7; Donner 2007.
14. Rivière and Licoppe 2005.
15. Rivière and Licoppe 2005:104.
16. Watkins 2007.
17. Ito 2005:1.
18. Ito 2005:4.
19. The rise of pagers in Japan is recounted by Okada 2005.
20. Matsuda 2005a:24. In personal email correspondence, Matsuda has explained that on other occasions, talking loudly in public is perfectly acceptable, such as at a barbeque in the park or during traditional cherry-blossom viewing. The subdued voices at Nara may have reflected the fact the park is a religious site.

21. Technical differences between Internet-based and non-Internet-based written messaging in Japan don't concern us here.

22. Okabe and Ito 2005:207.

23. Okabe and Ito 2005:208.

24. Okada 2005:49.

25. My special thanks to Misa Matsuda for calculating this subset of the larger Japanese corpus, which encompassed ages 12–69.

26. Dobashi 2005:228.

27. Matsuda 2005b:135.

28. Habuchi 2005:171.

29. Kato 2005:105.

30. Perry 1977:75.

31. http://www.itu.int/ITU-D/icteye/Indicators/Indicators.aspx#. Accessed June 23, 2007. Historical statistics continue to be updated on the site—hence the relevance of accession date.

32. Madden 2006.

33. Fox and Madden 2005.

34. Lenhart et al. 2005.

35. Lenhart et al. 2005.

36. Ling and Haddon In press.

37. Traugott et al. 2006.

38. M:Metrics 2006.

39. For details of the study, see Baron and Ling 2007.

40. M. Gergen 2005.

41. FasPay Technologies 2006.

42. Johnson 2006.

43. Again, my thanks to Misa Matsuda for calculating these statistics by gender.

44. Katz and Sugiyama 2005.

45. In Europe, seriatim messages can now be linked together, with the recipient seeing just one long message.

46. http://www.Internetworldstats.com/stats.htm. (Data posted as of February 2, 2007.)

47. Chemin and Malingre 2005.

48. Rich Ling personal communication.

49. Recall that in 2005, 75 percent of American online teens were IM users (Lenhart et al. 2005).

50. Döring 2002; Hård af Segerstad 2002; Ling 2005; Thurlow and Brown 2003.

51. For details on our findings, including levels of statistical significance, see Ling and Baron 2007.

52. Hård af Segerstad 2002.

53. Squires 2007.

54. Ling 2007.

55. In reporting their results, Thurlow and Brown don't distinguish between apostrophes used in contractions versus possessives.

Chapter 8

1. Sheidlower 1996:112–113. See Crystal 2007 for a contemporary discussion of linguistic prescriptivism.

2. Thurlow 2006.

3. These quotations correspond to article numbers 45, 24, 29, 20, 57, 40, and 79 in Thurlow 2006.

4. See Meyrowitz 1985 for an earlier discussion of American infatuation with youth culture.

5. http://homepage.ntlworld,com/vivian.c/Punctuation/ApostGrocers.htm.

6. See Lyons 1970 for an overview of Chomsky's relevant work.

7. Sapir 1921:39.

8. Saulny 2002:A1.

9. See Rosenberg 1995.

10. For discussion of this transformation see Baron 2000, chapter 5.

11. See, for example, Kakutani 2002.

12. "A foolish consistency is the hobgoblin of little minds, adored by little statesmen and philosophers and divines. With consistency a great soul has simply nothing to do." (Ralph Waldo Emerson, *Essays*, "Self-Reliance.")

13. Baron 2000.

14. Dave Barry, "Wit's End," *Washington Post Magazine*, March 4, 2001, 32.

15. Hale and Scanlon 1999:3, 9, 12, 15.

16. *BBC Online*, May 5, 2000.

17. "A Marriage by Telegraph" 1874.

18. "The Telegraph and the Mails" 1874.

19. See, for example, Scott and Jarvogen 1868; Gray 1885.

20. Lubrano 1997:124.

21. Gitelman 1999.

22. Koenigsberg 1987; Grimes 1992.

23. Fenwick 1948:14, 360.

24. Martin 1997:4.

25. Shea 1994:24.

26. J. Cohen 2000.

27. Research by Ylva Hård af Segerstad and Sylvana Sofkova Hashemi (2006) confirms that Swedish children between the ages of ten and fifteen have a good sense of differences between online and offline writing genres.

28. This example is taken from Cook 2004:94.

29. Hale and Scanlon 1999:95.

30. See Metcalf 2002:2–4, 44–51.

Chapter 9

1. Selgin 2003.

2. The original cartoon appeared in 1976 in the *Chronicle of Higher Education*. William Hamilton graciously redrew the cartoon for me in spring 2007 when we could not locate the original artwork.

3. Harris 1989.

4. Danet and Bogoch 1992.

5. Parkes 1991; Clanchy 1993.

6. Burrow, 1982:47.

7. Kastan 2001. Challenging this position, Lukas Erne (2003) argues that Shakespeare wrote different versions of some plays for publication than for the stage.

8. Chartier 1989; Transactions of the Book: A Conference at the Folger Shakespeare Library, Washington, DC, 2001.

9. Chadwick 1959.

10. Lipking 1998; J. W. Saunders 1951.

11. Ezell 1999.

12. Hornbeak 1934; Robertson 1942.

13. Gottlieb 2000.

14. Baron 2000.

15. Minnis 1988.

16. Jaszi 1991; Rose 1993; Woodmansee 1984; Woodmansee and Jaszi 1994.

17. Letter CXXIV, November 19, 1750. In Chesterfield 1901:355.

18. Parker 2001; Tufte 2003.

19. Early American copyright law largely derives from England.

20. See Turner 2006; Taylor and Harper 2002.

21. http://www.gnu.org/licenses/gpl-howto.html. Textual emphasis is in the GNU License.

22. See http://creativecommons.org/about for details on the history of the Creative Commons and how its licenses work. For discussion of why intellectual property should be placed in the public domain, see Lessig 2001.

23. Digital Dilemma 2000:133.

24. http://www.plos.org.

25. Barlow 1997:362.

26. New York Times Book Review, June 3, 2007, 4.

27. McCrum et al. 1986.

28. Gleick 1999.

29. Tannen 2003.

30. Two classic works on the development of clocks in the West are Carlo Cipolla's Clocks and Culture (1978) and David Landes's Revolution in Time (1983).

31. Eriksen 2001:43.

32. Carey 1983.

33. See Clark Blaise's Time Lord (2000).

34. Lubrano 1997:120.

35. Hall 1914:53–54.

36. Nelson 1980.

37. Ullman 1960:11; Rodriguez and Cannon 1994.

38. W. M. Saunders 1921:x–xi.

39. See John and Dianne Tillotson's web site on medieval writing, available at http://medievalwriting.50megs.com/writing.htm.

40. Havelock 1963.

41. Goody and Watt 1963:344.

42. Eisenstein 1979.

43. For background on the commonplace book tradition, see A. Moss 1996.

44. Baker 2000:8.

45. Eriksen 2001:59.

46. Eriksen 2001:67.

47. Rimer 2007.

48. Mood, like tense, is an attribute of verb conjugation in many Indo-European languages. The most familiar verb moods are indicative ("Jennifer will play the harp this afternoon") and subjunctive ("Jennifer would have played the harp this afternoon but Jeanne-marie could not accompany her").

49. For the story, see Plotz 2002.

50. The number of bachelor's degrees awarded in the United States rose from almost 27,500 in 1899–1900 to nearly 1,400,000 in 2003–2004 (National Center for Educational Statistics).

51. *Wall Street Journal*, March 15, 2007, D1.

52. May 22, 2006, news release of the Book Industry Study Group. Available at http://www.bisg.org/news/press.php?pressid=35.

53. See Table IV.5 Book production: number of titles by UDC classes. Available at http://www.uis.unesco.org/TEMPLATE/html/CultAndCom/Table_IV_5_America .html.

54. May 9, 2006, news release from Bowker, the publisher of *Books in Print*. Available at http://www.bowker.com/press/bowker/2006_0509_bowker.htm.

55. Amory 1996:51.

56. Reports can be found at http://nationsreportcard.gov.

57. http://nces.ed.gov/nationsreportcard/pdf/main2005/2007468.pdf. While scores for Hispanic students (a population that doubled between 1992 and 2005) remained largely the same over time, results for both non-Hispanic whites and African American students dropped significantly.

58. Hurwitz and Hurwitz 2004. The NAEP report was issued in July 2003.

59. Hurwitz and Hurwitz 2004.

60. Kaiser Family Foundation 2005:38.

61. The full report is available at http://nces.ed.gov/pubsearch/pubsinfo.asp? pubid=2007467.

62. The full NEA report is available at http://www.nea.gov/news/news04/Reading AtRisk.html.

63. The National Assessment of Adult Literacy is available at http://nces.ed.gov/ naal/pdf/2007464.pdf.

64. Address at the Phi Beta Kappa 41st Triennial Council Meeting, Atlanta, GA, October 25–29, 2006.

65. Weeks 2001; Zernike 2002.

66. Zernike 2002.

67. Bruns 1980:113.

68. For discussion of the invention of the phonograph, see Gitelman 1999 and the Library of Congress's American Memory site, http://memory.loc.gov/ammem/edhtml/edcyldr.html.

69. Hubert 1889:259, 260, 259.

70. Bellamy 1889:743, 744.

71. See, for example, Weedon 2003; Lehmann-Haupt 1951; Kaestle et al. 1991.

72. Vershbow 2006.

73. Mitchell 1995:56.

74. Negroponte 1995:164.

75. In fact, a growing number of publishers are successfully shifting their editorial processes—for both copyeditors and authors—from hardcopy to online.

Chapter 10

1. See Noguchi 2006 for discussion of location tracking services on mobile phones.

2. "US Highway Deaths on the Rise" 2006.

3. September 22, 2006.

4. Samuelson 2006.

5. Kinsley 2006.

6. *Time,* January 1, 2007.

7. Bell 2005:84–85.

8. Eriksen 2001:60.

9. Riesman 1950:v. Sherry Turkle (in press) argues that a fundamental consequence of being "always on" such information communication technologies as computers, the BlackBerry, and mobile phones is that we redefine our notion of individual self as being a product of our lives as lived on these technologies.

10. Avery et al. 1978:5.

11. Heffernan and Zeller 2006.

12. Conlin 2007:42.

13. D'Ausilio 2005.

14. Dux et al. 2006:1109.

15. Lohr 2007.

16. Iqbal and Horvitz 2007.

17. An audio file of Linda Stone's address to the Emerging Technology Conference in San Diego in March 2006 is available at http://www.itconversations.com/shows/detail739.html.

18. S. Levy 2006.

19. Cited by Freedman 2007.

20. Greenfield 2003.

21. Christakis et al. 2004.

22. Richtel 2007a.

23. Friedman 2006.

24. Cassidy 2006.

25. Barry 2007:13.

26. Markus 1994:119.

27. Kraut et al. 1998.

28. Nie and Hillygus 2002.

29. See, for example, Wellman et al. 2001; Quan-Haase and Wellman 2002.

30. Franzen 2000:435.

31. Kraut et al. 2002:49.

32. Veenhof 2006.

33. McQuigge 2006.

34. McPherson et al. 2006.

35. Eng et al. 2002.

36. Boase et al. 2006.

37. The Pew study speaks of "core" versus "significant" ties, rather than "strong" versus "weak."

38. Wright 2007.

39. Rheingold 1999.

40. Umble 1996.

41. Carr 2005.

42. Bugeja 2005:41.

43. Gardner 2006.

44. Bellamy 1889:743.

45. C. Levy 2007.

46. Okada 2002, cited in Matsuda 2005b:130.

47. Matsuda 2005b:137.

48. B. Stone 2007a.

49. Wurtzel and Turner 1977:49–50.

50. Wurtzel and Turner 1977:53–54.

51. Brady 2006.

52. "Some Tech-Gen Youth Go Offline" 2006.

53. Friedman 2006.

54. In May 2007, the state of Washington signed into law the first U.S. ban on driving while texting on a mobile phone. See Richtel 2007b.

55. "Schools Say iPods Become Tool for Cheaters" 2007.

56. N. Cohen 2007.

57. Lewis 1988.

58. "Some Tech-Gen Youth Go Offline" 2006.

59. Bruce 2007.

60. Manguel 1996:193. Manguel's source is Edward G. Browne's *A Literary History of Persia* (1959).

61. R. Moss 1996.

References

Abbate, Janet. 1999. *Inventing the Internet*. Cambridge, MA: MIT Press.

Acquisti, Alessandro, and Ralph Gross. 2006. "Imagined Communities: Awareness, Information Sharing, and Privacy on the Facebook." Sixth Workshop on Privacy Enhancing Technologies. Robinson College, Cambridge University. Cambridge, UK.

Adamczyk, Piotr, and Brian Bailey. 2004. "If Not Now, When?: The Effects of Interruption at Different Moments within Task Execution." *Proceedings of the SIGCHI Conference on Human Factors in Computing Systems* (CHI '04). Vienna, Austria, April 24–29. New York: ACM Press, 271–278.

Agar, Jon. 2003. *Constant Touch: A Global History of the Mobile Phone*. Duxford, Cambridge: Icon Books Ltd.

"A Marriage by Telegraph." 1874. *Telegrapher* 7 (May 1): 135.

Amory, Hugh. 1996. "The Trout and the Milk: An Ethnobibliographical Talk." *Harvard Library Bulletin* 7 (1): 50–65.

Annenberg Public Policy Center. 1996. "Call-In Political Talk Radio: Background, Content, Audiences, Portrayal in Mainstream Media." The Annenberg Public Policy Center of the University of Pennsylvania, August 7. Available at http://annenberg publicpolicycenter.org/02_reports_releases/report_1996.htm.

Argamon, Shlomo, Moshe Koppel, Jonathan Fine, and Anat Rachel Shimoni. 2003. "Gender, Genre, and Writing Style in Formal Written Texts." *Text* 23:321–346.

Armstrong, Cameron B., and Alan M. Rubin. 1989. "Talk Radio as Interpersonal Communication." *Journal of Communication* 39 (2): 84–94.

Armstrong, G. Blake, and Pradeep Sopory. 1997. "Effects of Background Television on Phonological and Visuo-Spatial Working Memory." *Communication Research* 24:459–480.

Avery, Robert K., Donald G. Ellis, and Thomas W. Glover. 1978. "Patterns of Communication on Talk Radio." *Journal of Broadcasting* 22 (1): 5–17.

Bahrampour, Tara. 2007. "On the Web, 'Dear Diary' Becomes 'Dear World': Teenagers Use Sites to Communicate, Vent." *Washington Post*, January 2, B2.

Bahrampour, Tara, and Lori Aratani. 2006. "Teens' Bold Blogs Alarm Area Schools." *Washington Post*, January 17, A1.

253

Baker, Nicholson. 2000. "Narrow Ruled." *American Scholar* 69 (Autumn): 5–8.

Barker, David C. 2002. *Rushed to Judgment: Talk Radio, Persuasion, and American Political Behavior.* New York: Columbia University Press.

Barker, Hannah. 1998. *Newspapers, Politics, and Public Opinion in Late Eighteenth-Century England.* Oxford: Clarendon Press.

———. 2002. "England, 1760–1815." In *Press, Politics, and the Public Sphere in Europe and North America, 1760–1820,* ed. H. Barker and S. Burrows, 93–112. Cambridge: Cambridge University Press.

Barker, Hannah, and Simon Burrows. 2002. "Introduction." In *Press, Politics, and the Public Sphere in Europe and North America, 1760–1820,* ed. H. Barker and S. Burrows, 1–22. Cambridge: Cambridge University Press.

Barlow, John Perry. 1997. "The Economy of Ideas: Everything You Know about Intellectual Property Is Wrong." In *Intellectual Property: Moral, Legal, and International Dilemmas,* ed. A. D. Moore, 349–371. Lanham, MD: Rowman & Littlefield.

Baron, Naomi S. 1998. "Letters by Phone or Speech by Other Means: The Linguistics of Email." *Language and Communication* 18 (2):133–170.

———. 2000. *Alphabet to Email: How Written English Evolved and Where It's Heading.* London: Routledge.

———. 2002. "Who Sets Email Style: Prescriptivism, Coping Strategies, and Democratizing Communication Access." *The Information Society* 18:403–413.

———. 2003. "Why Email Looks Like Speech: Proofreading, Pedagogy, and Public Face." In *New Media Language,* ed. J. Aitchison and D. Lewis, 102–113. London: Routledge.

———. 2004. " 'See You Online': Gender Issues in College Student Use of Instant Messaging." *Journal of Language and Social Psychology* 23 (4):397–423.

———. In press. "Discourse Structures in Instant Messaging: The Case of Utterance Breaks." In *Computer-Mediated Conversation,* ed. S. Herring. Cresskill, NJ: Hampton Press.

Baron, Naomi S., and Rich Ling. 2007. "Emerging Patterns of American Mobile Phone Use: Electronically-Mediated Communication in Transition." In *Mobile Media 2007: Proceedings of an International Conference on Social and Cultural Aspects of Mobile Phones, Media and Wireless Technologies,* June 2–4, ed. G. Goggin and L. Hjorth, 218–230. Sydney, Australia: University of Sydney.

Barry, Dave. 2007. "You've Got Trouble." Review of *Send: The Essential Guide to Email for Office and Home* by David Shipley and Will Schwalbe. Alfred Knopf. *New York Review of Books,* May 6, 13.

Baym, Nancy. 1995. "The Performance of Humor in Computer-Mediated Communication." *Journal of Computer-Mediated Communication* 1 (2). Available at http://www.ascusc.org/jcmc/vol1/issue2/baym.html.

Baym, Nancy, Yan Bing Zhang, and Mei-Chen Lin. 2004. "Social Interactions across Media: Interpersonal Communication on the Internet, Face-to-Face, and the Telephone." *New Media & Society* 6:299–318.

Beck, John, and Mitchell Wade. 2003. *DoCoMo: Japan's Wireless Tsunami.* New York: American Management Association.

Bell, Genevieve. 2005. "The Age of the Thumb: A Cultural Reading of Mobile Technologies from Asia." In *Thumb Culture: The Meaning of Mobile Phones for Society*, ed. P. Glotz, S. Bertschi, and C. Locke, 67–87. New Brunswick, NJ: Transaction Publishers.

Bellamy, Edward. 1889. "With the Eyes Shut." *Harper's New Monthly Magazine* 79 (473): 736–745.

Biber, Douglas. 1988. *Variation across Speech and Writing*. Cambridge: Cambridge University Press.

Biber, Douglas, Susan Conrad, and Randi Reppen. 1998. *Corpus Linguistics: Investigating Language Structure and Use*. Cambridge: Cambridge University Press.

Biber, Douglas, and Edward Finegan. 1997. "Diachronic Relations among Speech-Based and Written Registers in English." In *To Explain the Present: Studies in the Changing English Language in Honour of Matti Rissanen*, ed. T. Nevalainen and L. Kahlas-Tarkka, 253–275. Helsinki: Modern Language Society.

Black, Jeremy. 1987. *The English Press in the Eighteenth Century*. Philadelphia: University of Pennsylvania Press.

Blaise, Clark. 2000. *Time Lord: Sir Sandford Fleming and the Creation of Standard Time*. New York: Pantheon Books.

Blom, Philipp. 2005. *Enlightening the World: Encyclopédie, The Book That Changed the Course of History*. New York: Palgrave Macmillan.

Blumberg, Stephen J., and Julian V. Luke. 2007. "Wireless Substitution: Early Release of Estimates Based on Data from the National Health Interview Survey, July–December 2006." National Center for Health Statistics, Centers for Disease Control and Prevention. Available at http://www.cdc.gov/nchs/data/nhis/earlyrelease/wireless200705.pdf.

Boase, Jeffrey, John B. Horrigan, Barry Wellman, and Lee Rainie. 2006. "The Strength of Internet Ties." Pew Internet & American Life Project, January 25. Available at http://www.pewinternet.org/pdfs/PIP_Internet_ties.pdf.

Boettinger, H. M. 1977. *The Telephone Book: Bell, Watson, Vail and American Life, 1876–1976*. Croton-on-Hudson, NY: Riverwood Publishers, Ltd.

Boneva, Bonka, and Robert Kraut. 2002. "Email, Gender, and Personal Relations." In *The Internet in Everyday Life*, ed. B. Wellman and C. Haythornthwaite, 372–403. Oxford: Blackwell.

Boneva, Bonka, Amy Quinn, Robert Kraut, Sara Kiesler, and Irina Shklovski. 2006. "Teenage Communication in the Instant Messaging Era." In *Computers, Phones, and the Internet*, ed. R. Kraut, M. Brynin, and S. Kiesler, 201–218. Oxford: Oxford University Press.

Bourne, H. R. Fox. 1966. *English Newspapers: Chapters in the History of Journalism*. New York: Russell & Russell.

boyd, danah. 2004. "Friendster and Publicly Articulated Social Networking." *Proceedings of the ACM Conference on Human Factors in Computing Systems* (CHI '04). Vienna, Austria, April 24–29. New York: ACM Press, 1279–1282.

Brady, Diane. 2006. "*#?@the E-Mail. Can We Talk?" *Business Week*, December 4, 109.

Brewer, Annie M. 1996. *Talk Shows and Hosts on Radio*. 4th ed. Whitefoord Press.

Brooks, Colin. 1991. "John Reeves and His Correspondents: A Contribution to the Study of British Loyalism, 1792–1793." In *Après 89: La Révolution modèle ou repoussoir*, ed. L. Domergue and G. Lamoine, 49–76. Toulouse.

Brooks, Lee R. 1968. "Spatial and Verbal Components of the Act of Recall." *Canadian Journal of Psychology* 22:349–367.

Browne, Edward G. 1959. *A Literary History of Persia*. 4 vols. Cambridge: Cambridge University Press.

Bruce, Chaddus. 2007. "CIA Gets in Your Face(book)." *Wired News*, January 24. Available at http://www.wired.com/news/technology/internet/1,72545–0.html.

Bruns, Gerald. 1980. "The Originality of Texts in a Manuscript Culture." *Comparative Literature* 32:113–129.

Bugeja, Michael. 2005. *Interpersonal Divide: The Search for Community in a Technological Age*. New York: Oxford University Press.

Burrow, J. A. 1982. *Medieval Writers and Their Work*. Oxford: Oxford University Press.

Cameron, Deborah. 1998. "Gender, Language, and Discourse: A Review Essay." *Signs: Journal of Women in Culture and Society* 23:945–973.

Campbell, Todd. 1998. "The First E-Mail Message." *PreText Magazine*. Available at http://pretext.com/mar98/features/story2.htm.

Carey, James W. 1983. "Technology and Ideology: The Case of the Telegraph." In *Prospects, The Annual of American Cultural Studies*, ed. J. Salzman, 8:303–325.

Carr, David. 2005. "Taken to a New Place, by a TV in the Palm." *New York Times*, December 18, Week in Review, 3.

Cassidy, John. 2006. "Me Media." *New Yorker*, May 15, 50–59.

Castells, Manuel, Mireia Fernández-Ardèvol, Jack Linchuan Qiu, and Araba Sey. 2007. *Mobile Communication and Society: A Global Perspective*. Cambridge, MA: MIT Press.

Chadwick, John. 1959. "A Prehistoric Bureaucracy." *Diogenes* 26:7–18.

Chafe, Wallace. 1980. "The Deployment of Consciousness in the Production of a Narrative." In *The Pear Stories: Cognitive, Cultural, and Linguistic Aspects of Narrative Production*, ed. W. Chafe, 9–50. Norwood, NJ: Ablex.

―――. 1994. *Discourse, Consciousness, and Time: The Flow and Displacement of Conscious Experience in Speaking and Writing*. Chicago: University of Chicago Press.

―――. 2001. "The Analysis of Discourse Flow." In *The Handbook of Discourse Analysis*, ed. D. Schriffrin, D. Tannen, and H. E. Hamilton, 673–688. Malden, MA: Blackwell.

Chafe, Wallace, and Jane Danielewicz. 1987. "Properties of Spoken and Written Language." In *Comprehending Oral and Written Language*, ed. R. Horowitz and S. J. Samuels, 83–113. San Diego: Academic Press.

Chafe, Wallace, and Deborah Tannen. 1987. "The Relation between Written and Spoken Language." *Annual Review of Anthropology* 16:383–407.

Chalaby, Jean K. 1998. *The Invention of Journalism*. New York: St. Martin's Press.

Chambers, J. K. 1992. "Linguistic Correlates of Gender and Sex." *English World-Wide* 13:173–218.

Chartier, Roger, ed. 1989. *The Culture of Print: Power and the Uses of Print in Early Modern Europe*. Trans. Lydia G. Cochrane. Princeton, NJ: Princeton University Press.

Chemin, Ariane, and Virginie Malingre. 2005. "MSN Messenger, la messagerie qui dévore les soirées des ados." *Le Monde*, 9 Janvier.

Chesterfield, Earl of. 1901. *Letters to His Son: On the Fine Art of Becoming a Man of the World and a Gentleman*, with topical headings and a special introduction by Oliver H. G. Leigh. Vol. 1. Washington, DC: M. Walter Dunne.

Chomsky, Noam. 1965. *Aspects of the Theory of Syntax*. Cambridge, MA: MIT Press.

Christakis, Dimitri A., Frederick J. Zimmerman, David L. DiGiuseppe, and Carolyn A. McCarty. 2004. "Early Television Exposure and Subsequent Attentional Problems in Children." *Pediatrics* 113 (4): 708–713.

Cipolla, Carlo. 1978. *Clocks and Culture, 1300–1700*. New York: W.W. Norton.

Clanchy, M. T. 1993. *From Memory to Written Record: England 1066–1307*. 2nd ed. Oxford: Blackwell.

Coates, Jennifer. 1993. *Women, Men, and Language*. 2nd ed. London: Longman.

Cohen, Joyce. 2000. "Sorry for Your Loss But Not That Sorry." *New York Times*, December 7, E1, E10.

Cohen, Noam. 2007. "A History Department Bans Citing Wikipedia as a Research Source." *New York Times*, February 21.

Collot, Milena, and Nancy Belmore. 1996. "Electronic Language: A New Variety of English." In *Computer-Mediated Communication: Linguistic, Social, and Cross-Cultural Perspectives*, ed. S. Herring, 13–28. Amsterdam: John Benjamins.

Conlin, Michelle. 2007. "Cheating—or Postmodern Learning?" *Business Week*, May 14, 42.

Cook, Vivian. 2004. *The English Writing System*. London: Arnold.

Crook, Charles, and David Barrowcliff. 2001. "Ubiquitous Computing on Campus: Patterns of Engagement by University Students." *International Journal of Human-Computer Interaction* 13:245–256.

Crystal, David. 1975. *The English Tone of Voice*. London: St. Martins.

———. 2001. *Language and the Internet*. Cambridge: Cambridge University Press.

———. 2007. *The Fight for English: How Language Pundits Ate, Shot, and Left*. New York: Oxford University Press.

Cutrell, Edward, Mary Czerwinski, and Eric Horvitz. 2001. "Notification, Disruption, and Memory: Effects of Messaging Interruptions on Memory and Performance." In *Human Computer Interaction. INTERACT '01*. ed. M. Hirose, 263–269. Tokyo: IOS Press for IFIP.

Dabbish, Laura, and Robert Kraut. 2004. "Controlling Interruptions: Awareness Displays and Social Motivation for Coordination. *Proceedings of the ACM Conference on Computer Supported Cooperative Work* (CSCW '04). Chicago, IL, November 6–10. New York: ACM Press, 182–191.

Damos, Diane L., ed. 1991. *Multiple-Task Performance*. London: Taylor & Francis.

Danet, Brenda. 2001. *Cyberplay: Community Online*. London: Berg.

Danet, Brenda, and Bryna Bogoch. 1992. "From Oral Ceremony to Written Document: The Transitional Language of Anglo-Saxon Wills." *Language and Communication* 12 (2): 95–122.

Daoussis, Leonard, and Stuart J. McKelvie. 1986. "Musical Preferences and Effects of Music on a Reading Comprehension Test for Extraverts and Introverts." *Perceptual and Motor Skills* 62:283–289.

D'Ausilio, Rosanne. 2005. "Multitasking Part II: De-Stressing." *TMCnet*, November 28. Available at http://www.tmcnet.com/news/2005/nov/1215030.htm.

Diderot, Denis, Jean D'Alembert, and a Society of Men of Letters. 1965. *Encyclopedia: Selections*. Trans., with an introduction and notes by Nelly S. Hoyt and Thomas Cassirer. Indianapolis: Bobbs-Merrill Company, Inc.

Digital Dilemma. 2000. National Research Council, Committee on Intellectual Property and the Emerging Information Infrastructure. Washington, DC: National Academy Press.

Dobashi, Shingo. 2005. "The Gendered Use of *Keitai* in Domestic Contexts." In *Personal, Portable, Pedestrian: Mobile Phones in Japanese Life*, ed. M. Ito, D. Okabe, and M. Matsuda, 219–236. Cambridge, MA: MIT Press.

Dodgson, Charles Lutwidge. 2001. *The Political Pamphlets and Letters of Charles Lutwidge Dodgson and Related Pieces*. Compiled, with introductory essays, notes, and annotations by Francine F. Abeles. Charlottesville, VA: Lewis Carroll Society of North America, distributed by the University Press of Virginia.

Donner, Jonathan. 2005. "The Social and Economic Implications of Mobile Telephony in Rwanda: An Ownership/Access Typology." In *Thumb Culture: The Meaning of Mobile Phones for Society*, ed. P. Glotz, S. Bertschi, and C. Locke, 37–51. New Brunswick, NJ: Transaction Publishers.

Donner, Jonathan. 2007. "The Rules of Beeping: Exchanging Messages via Intentional 'Missed Calls' on Mobile Phones." *Journal of Computer-Mediated Communication* 13 (1). Available at http://jcmc.indiana.edu/vol13/issue1/donner.html.

Döring, Nicola. 2002. " 'Kurzm. wird gesendet'—Abkürzungen und Akronyme in der SMS-Kommunikation." *Muttersprache. Vierteljahresschrift für deustsche Sprache*, 2.

Doyle, Arthur Conan. 1986. *Letters to the Press*. Compiled, with an introduction by John Michael Gibson and Richard Lancelyn Green. London: Secker & Warburg.

Draganski, Bogdan, Christian Gaser, Volker Busch, Gerhard Schuierer, Ulrich Bogdahn, and Arne May. 2004. "Changes in Grey Matter Induced by Training." *Nature* 427 (January 22): 311–312.

Dux, Paul E., Jason Ivanoff, Christopher L. Asplund, and René Marois. 2006. "Isolation of a Central Bottleneck of Information Processing with Time-Resolved fMRI." *Neuron* 52:1109–1120.

Eckert, Penelope, and Sally McConnell-Ginet. 2003. *Language and Gender*. New York: Cambridge University Press.

Einbinder, Harvey. 1964. *The Myth of the Britannica*. New York: Grove Press.

Eisenstein, Elizabeth. 1979. *The Printing Press as an Agent of Change*. Cambridge: Cambridge University Press.

Ellison, Nicole, Charles Steinfield, and Cliff Lampe. 2007. "The Benefits of Facebook 'Friends': Social Capital and College Students' Use of Online Social Network Sites." *Journal of Computer-Mediated Communication*, 12(4). Available at http://jcmc.indiana.edu/vol12/issue4//ellison.html.

Emigh, William, and Susan C. Herring. 2005. "Collaborative Authoring on the Web: A Genre Analysis of Online Encyclopedias." *Proceedings of the Thirty-Eighth Hawai'i International Conference on System Sciences* (HICSS-38). Los Alamitos: IEEE Press.

Encyclopaedia Britannica, Inc. 2006. "Fatally Flawed: Refuting the Recent Study on Encyclopedic Accuracy by the Journal *Nature*." March. Available at http://corporate.britannica.com/britannica_nature_response.pdf.

Eng, Patricia M., Eric B. Rimm, Garrett Fitzmaurice, and Ichiro Kawachi. 2002. "Social Ties and Change in Social Ties in Relation to Subsequent Total and Cause-Specific Mortality and Coronary Heart Disease Incidence in Men." *American Journal of Epidemiology* 155 (8): 700–709.

Erasmus, Desiderius. 1569. *De copia verborum ac rerum (On Copia of Words and Ideas)*. Trans. Donald B. King and H. David Rix. 1963. Milwaukee, WI: Marquette University Press.

Eriksen, Thomas H. 2001. *Tyranny of the Moment: Fast and Slow Time in the Information Age*. London: Pluto Press.

Erne, Lukas. 2003. *Shakespeare as Literary Dramatist*. Cambridge: Cambridge University Press.

Ezell, Margaret. 1999. *Social Authorship and the Advent of Print*. Baltimore: Johns Hopkins University Press.

FasPay Technologies. 2006. "Billboard to Chart Ringtones." Available at http://faspay.com/site/print.php?pid=Billboard_to_chart_ringtones.

Fenwick, Millicent. 1948. *Vogue's Book of Etiquette*. New York: Simon and Schuster.

Ferrara, Kathleen, Hans Brunner, and Greg Whitemore. 1991. "Interactive Written Discourse as an Emergent Register." *Written Communication* 8:8–34.

Feuer, Alan, and Jason George. 2005. "Internet Fame is Cruel Mistress for a Dancer of the Numa Numa." *New York Times*, February 26.

Finder, Alan. 2006. "When a Risqué Online Persona Undermines a Chance for a Job." *New York Times*, June 11.

Fine, Jon. 2006. "Polluting the Blogosphere." *Business Week*, July 10, 20.

Floro, Maria Sagrario, and Marjorie Miles. 2003. "Time Use, Work and Overlapping Activities: Evidence from Australia." *Cambridge Journal of Economics* 27:881–904.

Fox, Susannah, and Mary Madden. 2005. "Generations Online: Data Memo." Pew Internet & American Life Project, December. Available at http://www.pewinternet.org/pdfs/PIP_Generations_Memo.pdf.

Fram, Alan. 2007. "Cellphone-Only Use Growing among Youths." Associated Press, May 14. Available at http://www.usatoday.com/tech/wireless/phones/2007–05–14-cellphone-only_N.htm.

Franzen, Axel. 2000. "Does the Internet Make Us Lonely?" *European Sociological Review* 16 (4): 427–438.

Freedman, David H. 2007. "What's Next: Taskus Interruptus." *Inc Magazine*, February. Available at http://www.inc.com/magazine/20070201/column-freedman.html.

Friedman, Thomas. 2006. "The Taxi Driver." Op-Ed, *New York Times*, November 1.

Gardner, Ralph Jr. 2006. "In College, You Can Go Home Again and Again." *New York Times*, December 14, E9.

Gatrell, V. A. C. 1994. *The Hanging Tree: Execution and the English People 1770–1868*. Oxford: Oxford University Press.

Gaver, William. 1991. "Technology Affordances." *Proceedings of the SIGCHI Conference on Human Factors in Computer Systems* (CHI '91). New Orleans, LA, April 28–June 5. New York: AMC Press, 79–84.

Gergen, Kenneth. 1991. *The Saturated Self: Dilemmas of Identity in Contemporary Life*. New York: Basic Books.

Gergen, Mary. 2005. "Using Mobile Phones: A Survey of College Women and Men." Paper presented at the 55th Annual International Communication Association, preconference session, "Mobile Communication: Current Research and Future Directions." New York, May 26.

Gibson, James J. 1979. *The Ecological Approach to Visual Perception*. New York: Houghton Mifflin.

Giles, Jim. 2005. "Internet Encyclopedias Go Head to Head." *Nature* 438 (December 15): 900–901.

Gitelman, Lisa. 1999. *Scripts, Grooves, and Writing Machines: Representing Technology in the Edison Era*. Stanford, CA: Stanford University Press.

Gleick, James. 1999. *Faster: The Acceleration of Just About Everything*. New York: Vintage Books.

Glotz, Peter, Stefan Bertschi, and Chris Locke, eds. 2005. *Thumb Culture: The Meaning of Mobile Phones for Society*. New Brunswick, NJ: Transaction Publishers.

Goffman, Erving. 1959. *The Presentation of Self in Everyday Life*. Garden City, NY: Doubleday.

Goggin, Gerard. 2006. *Cell Phone Culture: Mobile Technology in Everyday Life*. New York: Routledge.

Golder, Scott, Dennis Wilkinson, and Bernardo A. Huberman. 2006. "Rhythms of Social Interaction: Messaging within a Massive Online Network." Information Dynamics Laboratory, HP Labs. Available at http://www.hpl.hp.com/research/idl/papers/facebook/index.html.

Gonzales, Suzannah. 2006. "Are You on the Facebook?" *Austin American-Statesman*, January 2.

Goody, Jack, and Ian Watt. 1963. "The Consequences of Literacy." *Comparative Studies in Society and History* 5 (3): 304–345.

Gottlieb, Nanette. 2000. *Word-Processing Technology in Japan: Kanji and the Keyboard*. Richmond, Surrey: Curzon.

Graf, Joseph. 2006. "The Audience for Political Blogs." Institute for Politics, Democracy, & the Internet, in collaboration with @dvocacy Inc. George Washington University Graduate School of Political Management, October.

Gray, Morris. 1885. *A Treatise on Communication by Telegraph*. Boston: Little, Brown, and Company.

Greenfield, Susan. 2003. *Tomorrow's People*. London: Allen Lane.

Grimes, William. 1992. "Great 'Hello' Mystery Is Solved." *New York Times*, March 5, C1.

Grinter, Rebecca, and Leysia Palen. 2002. "Instant Messaging in Teen Life." *Proceedings of the ACM Conference on Computer Supported Cooperative Work* (CSCW '02). New Orleans, LA, November 16–20. New York: ACM Press, 21–30.

Gross, Ralph, and Alessandro Acquisti. 2005. "Information Revelation and Privacy in Online Social Networks: The Facebook Case." ACM Workshop on Privacy in the Electronic Society. Alexandria, VA, November 7.

GSM World. Available at http://www.gsmworld.com.

Habuchi, Ichiyo. 2005. "Accelerating Reflexivity." In *Personal, Portable, Pedestrian: Mobile Phones in Japanese Life*, ed. M. Ito, D. Okabe, and M. Matsuda, 163–182. Cambridge, MA: MIT Press.

Hale, Constance, and Scanlon, Jessie. 1999. *Wired Style*. Rev. ed. New York: Broadway Books.

Hall, Florence Howe. 1914. *Good Form for All Occasions*. New York: Harper & Brothers.

Halliday, M. A. K. 1967. "Notes on Transitivity and Theme in English, Part 2." *Journal of Linguistics* 3:199–244.

Hallowell, Edward M. 2006. *CrazyBusy: Overstretched, Overbooked, and About to Snap! Strategies for Coping in a World Gone ADD*. New York: Ballantine Books.

Hamill, Lynne, and Amparo Lasen, eds. 2005. *Mobile World: Past, Present, and Future*. London: Springer.

Hård af Segerstad, Ylva. 2002. *Use and Adaptation of Written Language to the Conditions of Computer-Mediated Communication*. Department of Linguistics, Göteborg, Sweden: Göteborg University.

Hård af Segerstad, Ylva, and Sylvana Sofkova Hashemi. 2006. "Learning to Write in the Information Age: A Case Study of Schoolchildren's Writing in Sweden." In *Writing and Digital Media*, eds. L. van Waes, M. Leijten, and C. Neuwirth, 49–64. Amsterdam: Elsevier.

Harper, Richard, Leysia Palen, and Alex Taylor, eds. 2005. *The Inside Text: Social, Cultural and Design Perspectives on SMS*. Dordrecht: Springer.

Harris, William. 1989. *Ancient Literacy*. Cambridge: Cambridge University Press.

Havelock, Eric. 1963. *Preface to Plato*. Cambridge, MA: Harvard University Press.

Heffernan, Virginia, and Tom Zeller, Jr. 2006. "The Lonelygirl That Really Wasn't." *New York Times*, September 13.

Hembrooke, Helene, and Geri Gay. 2003. "The Laptop and the Lecture: The Effects of Multitasking in Learning Environments." *Journal of Computing in Higher Education* 15:46–64.

Herring, Susan, ed. 1996. *Computer-Mediated Communication: Linguistic, Social, and Cross-Cultural Perspectives*. Amsterdam: John Benjamins.

———. 2000. "Gender Differences in CMC: Findings and Implications." *The CPSR Newsletter*, Winter. Available at http://www.cpsr.org/publications/newsletters/issues/2000/Winter2000/herring.html.

————. 2003. "Gender and Power in Online Communication." In *The Handbook of Language and Gender*, ed. J. Holmes and M. Meyerhoff, 202–228. Oxford: Blackwell.

Herring, Susan, Inna Kouper, Lois Ann Scheidt, and Elijah Wright. 2004. "Women and Children Last: The Discursive Construction of Weblogs." In *Into the Blogosphere: Rhetoric, Community, and Culture of Weblogs*, ed. L. J. Gurak, S. Antonijevic, L. Johnson, C. Ratliff, and J. Reyman. Available at http://blog.lib.umn.edu/blogosphere/women_and_children.html.

Herring, Susan, Inna Kouper, John Paolillo, Lois Ann Scheidt, Michael Tyworth, Peter Welsch, Elijah Wright, and Ning Yu. 2005. "Conversations in the Blogosphere: An Analysis 'From the Bottom Up.'" *Proceedings of the Thirty-Eighth Hawai'i International Conference on System Sciences* (HICSS-38). Los Alamitos: IEEE Press.

Herring, Susan, Lois Ann Scheidt, Sabrina Bonus, and Elijah Wright. 2005. "Weblogs as a Bridging Genre." *Information, Technology & People* 18 (2): 142–171.

Herring, Susan, and John Paolillo. 2006. "Gender and Genre Variation in Weblogs." *Journal of Sociolinguistics* 10 (4): 439–459.

Hewlett-Packard. 2005. "Abuse of Technology Can Reduce UK Workers' Intelligence." *Small & Medium Business* press release, April 22.

Holmes, Janet. 1993. "Women's Talk: The Question of Sociolinguistic Universals." *Australian Journal of Communication* 20:125–149.

————. 1995. *Women, Men, and Politeness*. New York: Longman.

Holmes, Janet, and Miriam Meyerhoff, eds. 2003. *The Handbook of Language and Gender*. Malden, MA: Blackwell.

Hornbeak, Katherine Gee. 1934. "The Complete Letter Writer in English, 1568–1800." *Smith College Studies in Modern Languages* 15 (3–4).

Horst, Heather, and Daniel Miller. 2006. *The Cell Phone: An Anthropology of Communication*. Oxford: Berg.

Hubert, Philip G., Jr. 1889. "The New Talking-Machines." *The Atlantic Monthly* 63 (376): 256–261.

Hurwitz, Nina, and Sol Hurwitz. 2004. "Words on Paper." *American School Board Journal* 191 (3): 16–20.

"IBM to Build Virtual Stores in Second Life." 2007. Associated Press, January 9. Available at http://www.cbc.ca/consumer/story/2007/01/09/tech-ibm.html.

International Telecommunication Union. 2007. ICT statistics database: Country data by region. Available at http://www.itu.int/ITU-D/icteye/Indicators/Indicators.aspx.

Iqbal, Shamsi T., and Eric Horvitz. 2007. "Disruption and Recovery of Computing Tasks: Field Study, Analysis, and Directions." *Proceedings of the ACM SIGCHI Conference on Human Factors in Computing Systems* (CHI '07). San Jose, CA, April 28–May 3. New York: ACM Press, 677–686.

Ironmonger, Duncan. 2003. "There are Only 24 Hours in a Day! Solving the Problematic of Simultaneous Time." *Proceedings of the 25th IATUR Conference on Time Use Research*. Brussels, Belgium.

Isaacs, Ellen, Alan Walendowski, Steve Whittaker, Diane Schiano, and Candace Kamm. 2002. "The Character, Functions, and Styles of Instant Messaging in the Workplace." *Proceedings of the Conference on Computer Supported Cooperative Work* (CSCW '02). New Orleans, LA, November 16–20. New York: ACM Press, 11–20.

Ito, Mizuko. 2005. "Introduction." In *Personal, Portable, Pedestrian: Mobile Phones in Japanese Life*, ed. M. Ito, D. Okabe, and M. Matsuda, 1–16. Cambridge, MA: MIT Press.

Ito, Mizuko, Daisuke Okabe, and Misa Matsuda, eds. 2005. *Personal, Portable, Pedestrian: Mobile Phones in Japanese Life*. Cambridge, MA: MIT Press.

Jacobs, Gloria. 2003. "Breaking Down Virtual Walls: Understanding the Real Space / Cyberspace Connections of Language and Literacy in Adolescents' Use of Instant Messaging." Paper presented at the American Educational Research Association, Chicago, IL, April 21–25.

James, Deborah. 1996. "Women, Men, and Prestige Speech Forms: A Critical Review." In *Rethinking Language and Gender Research*, ed. V. L. Bergvall, J. M. Bing, and A. F. Freed, 98–125. London: Longman.

Jaszi, Peter. 1991. "Toward a Theory of Copyright: The Metamorphoses of 'Authorship.'" *Duke Law Journal*, April, 455–502.

Johnson, Carolyn. 2006. "Do u txt ur kdz?" *Boston Globe*, December 17.

Kaestle, Carl F., Helen Damon-Moore, Lawrence C. Stedman, Katherine Tinsley, and William V. Trollinger, Jr. 1991. *Literacy in the United States: Readers and Reading since 1880*. New Haven, CT: Yale University Press.

Kaiser Family Foundation. 2005. "Generation M: Media in the Lives of 8–18 Year-Olds." Available at http://www.kff.org/entmedia/upload/Executive-Summary-Generation-M-Media-in-the-Lives-of-8–18-Year-olds.pdf.

Kakutani, Michiko. 2002. "Debate? Dissent? Discussion? Oh, Don't Go There!" *New York Times*, March 23.

Kalman, Yoram M., Gilad Ravid, Daphne R. Raban, and Sheizaf Rafaeli. 2006. "Pauses and Response Latencies: A Chronemic Analysis of Asynchronous CMC." *Journal of Computer-Mediated Communication* 12 (1). Available at http://jcmc .indiana.edu/vol12/issue1/kalman.html.

Kastan, David S. 2001. *Shakespeare and the Book*. Cambridge: Cambridge University Press.

Kato, Haruhiro. 2005. "Japanese Youth and the Imagining of *Keitai*." In *Personal, Portable, Pedestrian: Mobile Phones in Japanese Life*, ed. M. Ito, D. Okabe, and M. Matsuda, 103–119. Cambridge, MA: MIT Press.

Katz, James E., ed. In press. *Handbook of Mobile Communication Studies*. Cambridge, MA: MIT Press.

Katz, James E., and Mark Aakhus, eds. 2002. *Perpetual Contact: Mobile Communication, Private Talk, Public Performance*. Cambridge: Cambridge University Press.

Katz, James E., and Ronald Rice. 2002. *Social Consequences of Internet Use: Access, Involvement, and Interaction*. Cambridge, MA: MIT Press.

Katz, James E., and Satomi Sugiyama. 2005. "Mobile Phones as a Fashion Statement." In *Mobile Communications: Re-Negotiation of the Social Sphere*, ed. R. Ling and P. Pedersen, 63–81. London: Springer.

Kendon, Adam. 1980. "The Sign Language of the Women of Yuendumu: A Preliminary Report on the Structure of Warlpiri Sign Language." *Sign Language Studies* 27:101–112.

Kenyon, Susan, and Glenn Lyons. 2007. "Introducing Multitasking to the Study of Travel and ICT." *Transportation Research A* 41 (2): 161–175.

Kinsley, Michael. 2006. "Like I Care." Op-Ed, *Washington Post*, November 28, A19.

Koenigsburg, Allen. 1987. "The First 'Hello!': Thomas Edison, the Phonograph, and the Telephone." *The Antique Phonograph Monthly* 8 (6).

Kogan, Herman. 1958. *The Great EB: The Story of the Encyclopedia Britannica.* Chicago: University of Chicago Press.

Kohiyama, Kenji. 2005. "A Decade in the Development of Mobile Communications in Japan (1993–2002)." In *Personal, Portable, Pedestrian: Mobile Phones in Japanese Life*, ed. M. Ito, D. Okabe, and M. Matsuda, 61–74. Cambridge, MA: MIT Press.

Koppel, Moshe, Shlomo Argamon, and Anat Rachel Shimoni. 2002. "Automatically Categorizing Written Texts by Author Gender." *Literary and Linguistic Computing* 17:401–412.

Kraut, Robert, Sara Kiesler, Bonka Boneva, Jonathon Cummings, Vicki Helgeson, and Anne Crawford. 2002. "Internet Paradox Revisited." *Journal of Social Issues* 58 (1): 49–74.

Kraut, Robert, Michael Patterson, Vicki Lundmark, Sara Kiesler, Tridas Mukopadhyay, and William Scherlis. 1998. "Internet Paradox: A Social Technology that Reduces Social Involvement and Psychological Well-Being?" *American Psychologist* 53 (9): 1017–1031.

Krishnamurthy, Sandeep. 2002. "The Multidimensionality of Blog Conversations: The Virtual Enactment of September 11." Paper presented at AOIR 3.0, Association of Internet Researchers, Maastricht, The Netherlands, October 13–16.

Labov, William. 1991. "The Intersection of Sex and Social Class in the Course of Linguistic Change." *Language Variation and Change* 2:205–254.

Landes, David S. 1983. *Revolution in Time: Clocks and the Making of the Modern World.* Cambridge, MA: Harvard University Press.

Lasica, J. D. 2001. "Blogging as a Form of Journalism." *USC Annenberg Online Journalism Review.* Available at http://www.ojr.org/ojr/workplace/1017958873.php.

Laufer, Peter. 1995. *Inside Talk Radio: America's Voice or Just Hot Air?* New York: Carol Publishing Group.

Lehmann-Haupt, Hellmut. 1951. *The Book in America: A History of the Making and Selling of Books in the United States.* 2nd ed. New York: Bowker.

Lenhart, Amanda, and Susannah Fox. 2006. "Bloggers: A Portrait of the Internet's New Storytellers." Pew Internet & American Life Project, July 19. Available at http://www.pewinternet.org/pdfs/PIP%20Bloggers%20Report%20July%2019%202006.pdf.

Lenhart, Amanda, Mary Madden, and Paul Hitlin. 2005. "Teens and Technology." Pew Internet & American Life Project, July 27. Available at http://www.pewinternet.org/pdfs/PIP_Teens_Tech_July2005web.pdf.

Lenhart, Amanda, Lee Rainie, and Oliver Lewis. 2001. "Teenage Life Online: The Rise of the Instant-Message Generation and the Internet's Impact on Friendships and Family Relationships." Pew Internet & American Life Project. Available at http://www.pewinternet.org/pdfs/PIP_Teens_Report.pdf.

Lessig, Lawrence. 2001. *The Future of Ideas: The Fate of the Commons in a Connected World*. New York: Random House.

Leuf, Bo, and Ward Cunningham. 2001. *The Wiki Way: Quick Collaboration on the Web*. Boston: Addison-Wesley.

Levy, Clifford J. 2007. "To Reach Me in Russia, Just Dial Brooklyn." *New York Times*, March 11, Week in Review.

Levy, David. 2001. *Scrolling Forward: Making Sense of Documents in the Digital Age*. New York: Arcade Publishing.

Levy, Steven. 2006. "Digital Distractions Bad for the Workplace." *Newsweek*, March 27. Available at http://www.msnbc.msn.com/id/11899893/site/newsweek/.

Lewis, Peter. 1988. "A Basic Choice with Laser Printers." *New York Times*, February 7, F12.

Ling, Rich. 2004. *The Mobile Connection: The Cell Phone's Impact on Society*. San Francisco, CA: Morgan Kaufmann.

———. 2005. "The Sociolinguistics of SMS: An Analysis of SMS Use by a Random Sample of Norwegians." In *Mobile Communications: Re-Negotiation of the Social Sphere*, ed. R. Ling and P. Pedersen, 335–349. London: Springer.

———. 2007. "The Length of Text Messages and Use of Predictive Texting: Who Uses It and How Much Do They Have to Say?" *AU TESOL Working Papers*, no. 4, American University, Washington, DC. Available at http://www.american.edu/tesol/WorkingPapers04.html.

Ling, Rich, and Naomi S. Baron. 2007. "Text Messaging and IM: A Linguistic Comparison of American College Data." *Journal of Language and Social Psychology* 26 (3): 291–298.

Ling, Rich, and Leslie Haddon. In press. "Mobile Emancipation: Children, Youth, and the Mobile telephone." In *International Handbook of Children, Media, and Culture*, ed. K. Dortner and S. Livingstone. London: Sage.

Ling, Rich, and Per Pedersen, eds. 2005. *Mobile Communications: Re-Negotiation of the Social Sphere*. London: Springer.

Lipking, Lawrence. 1998. *Samuel Johnson: The Life of an Author*. Cambridge, MA: Harvard University Press.

Livingstone, Sonia, and Peter Lunt. 1994. *Talk on Television: Audience Participation and Public Debate*. London: Routledge.

Locke, John L. 1998. *The De-Voicing of Society: Why We Don't Talk to Each Other Anymore*. New York: Simon & Schuster.

Lohr, Steve. 2007. "Slow Down, Multitaskers, and Don't Read in Traffic." *New York Times*, March 25.

Lubrano, Annteresa. 1997. *The Telegraph: How Technology Innovation Caused Social Change*. New York: Garland.

Lyons, John. 1970. *Noam Chomsky*. New York: Viking.

Madden, Mary. 2006. "Internet Penetration and Impact: Data Memo." Pew Internet & American Life Project, April. Available at http://www.pewinternet.org/pdfs/PIP_Internet_Impact.pdf.

Maguire, Eleanor A., David G. Gadian, Ingrid S. Johnsrude, Catriona D. Good, John Ashburner, Richard S. J. Frackowiak, and Christopher D. Frith. 2000. "Navigation-Related Structure Change in the Hippocampi of Taxi Drivers." *Proceedings of the National Academy of Sciences* 97 (April 11): 4398–4403.

Manguel, Alberto. 1996. *A History of Reading.* New York: Viking.

Manhart, Klaus. 2004. "The Limits of Multitasking." *Scientific American Mind* 1 (December): 62–67.

Markus, M. Lynne. 1994. "Finding a Happy Medium: Explaining the Negative Effects of Electronic Communication on Social Life at Work." *ACM Transactions on Information Systems* 12 (1): 119–149.

Martin, Judith. 1997. *Miss Manners' Basic Training: Communication.* New York: Crown.

Matsuda, Misa. 2005a. "Discourses of *Keitai* in Japan." In *Personal, Portable, Pedestrian: Mobile Phones in Japanese Life*, ed. M. Ito, D. Okabe, and M. Matsuda, 19–39. Cambridge, MA: MIT Press.

———. 2005b. "Mobile Communication and Selective Sociality." In *Personal, Portable, Pedestrian: Mobile Phones in Japanese Life*, ed. M. Ito, D. Okabe, and M. Matsuda, 123–142. Cambridge, MA: MIT Press.

Maynor, Natalie. 1994. "The Language of Electronic Mail: Written Speech." In *Contemporary Usage Studies*, ed. G. Little and M. Montgomery, 48–54. American Dialect Society Publications. Tuscaloosa: University of Alabama Press.

McCrum, Robert, William Cran, and Robert MacNeil. 1986. *The Story of English.* New York: Viking.

McHenry, Robert. 2003. "Whatever Happened to Encyclopedic Style?" *Chronicle of Higher Education* 49 (February 28): B13–14.

———. 2004. "The Faith-Based Encyclopedia." *TSC Daily*, November 15. Available at http://www.tcsdaily.com/printArticle.aspx?ID=111504A.

McNeill, Laurie. 2003. "Teaching an Old Genre New Tricks: The Diary on the Internet." *Biography: An Interdisciplinary Quarterly* 26:24–47.

McPherson, Miller, Lynn Smith-Lovin, and Matthew E. Brashears. 2006. "Social Isolation in America: Changes in Core Discussion Networks over Two Decades." *American Sociological Review* 71 (3): 353–375.

McQuigge, Michelle. 2006. "Heavy Internet Users Spend Less Time with Family, Friends: Statistics Canada." *Canoe Network*, August 2. Available at http://money.canoe.ca/News/Sectors/Entertainment/2006/08/02/pf-1715240.html.

McWhorter, John. 2003. *Doing Our Own Thing: The Degradation of Language and Music and Why We Should, Like, Care.* New York: Gotham Books.

Metcalf, Allan. 2002. *Predicting New Words.* Boston: Houghton Mifflin.

Meyrowitz, Joshua. 1985. *No Sense of Place: The Impact of Electronic Media on Social Behavior.* New York: Oxford University Press.

Michelson, William. 2005. *Time Use: Expanding Explanation in the Social Sciences.* Boulder: Paradigm Publishers.

Milgram, Stanley. 1970. "The Experience of Living in Cities." *Science* 167 (March): 1461–1468.

Minnis, A. J. 1988. *Medieval Theories of Authorship*. 2nd ed. Philadelphia: University of Pennsylvania.

Mitchell, William. 1995. *City of Bits: Space, Place, and the Infobahn*. Cambridge, MA: MIT Press.

M:Metrics. 2006. "Teens Take User-Generated Content and Social Networking to Go," *M:Metrics* press release, December 14. Available at http://www.mmetrics .com/press/PressRelease.aspx?article=20061214-social-networking.

Morton, David. 2000. *Off the Record: The Technology and Culture of Sound Recording in America*. New Brunswick, NJ: Rutgers University Press.

Moss, Ann. 1996. *Printed Commonplace-Books and the Structuring of Renaissance Thought*. Oxford: Clarendon Press.

Moss, Rose. 1996. "No Space Like Shared Space." *New York Times*, January 15, A17.

Mulac, Anthony, and Torborg Louisa Lundell. 1994. "Effects of Gender-Linked Language Differences in Adults' Written Discourse: Multivariate Tests of Language Effects." *Language and Communication* 14 (3): 299–309.

Munson, Wayne. 1993. *All Talk: The Talkshow in Media Culture*. Philadelphia: Temple University Press.

Musgrove, Mike. 2007. "E-Mail Reply to All: 'Leave Me Alone.'" *Washington Post*, May 25, A1.

Nardi, Bonnie, Steve Whittaker, and Erin Bradner. 2000. "Interaction and Outeraction: Instant Messaging in Action." *Proceedings of the ACM Conference on Computer Supported Cooperative Work* (CSCW '00). Philadelphia, PA, December 2–6. New York: ACM Press, 79–88.

Nardi, Bonnie, Diane J. Schiano, and Michelle Gumbrecht. 2004. "Blogging as Social Activity, or, Would You Let 900 Million People Read Your Diary?" *Proceedings of the ACM Conference on Computer Supported Cooperative Work* (CSCW '04). Chicago, IL, November 6–10. New York: ACM Press, 222–231.

National Center for Educational Statistics. 2002. *The Nation's Report Card: Writing 2002*. National Assessment of Educational Progress. Washington, DC: U.S. Department of Education, Institute of Educational Sciences.

National Center for Educational Statistics. Digest of Educational Statistics: 2005. Table 169. Historical summary of faculty, students, degrees, and finances in degree-granting institutions: Selected years, 1869–70 through 2003–04. Available at http://nces.ed.gov/programs/digest/d05/tables/dt05_169.asp.

Negroponte, Nicholas. 1995. *Being Digital*. New York: Knopf.

Nelson, Daniel. 1980. *Frederick W. Taylor and the Rise of Scientific Management*. Madison: University of Wisconsin Press.

Newitz, Annalee. 2006. "Your Second Life Is Ready." *Popular Science* 269 (3): 75–98.

Nie, Norman H., and D. Sunshine Hillygus. 2002. "The Impact of Internet Use on Sociability: Time-Diary Findings." *IT& Society* 1(1): 1–20. Available at http://www .ITandSociety.org.

Noguchi, Yuki. 2006. "Friends at Hand and In Your Face." *Washington Post*, December 29, D1.

Okabe, Daisuke, and Mizuko Ito. 2005. "*Keitai* in Public Transportation." In *Personal, Portable, Pedestrian: Mobile Phones in Japanese Life*, ed. M. Ito, D. Okabe, and M. Matsuda, 205–217. Cambridge, MA: MIT Press.

Okada, Tomoyuki. 2002. "Keitai kara Manabu to Iukoto (What It Means to Learn from *Keitai*)." In *Keitai-Gaku Nyumon (Understanding Mobile Media)*. ed. T. Okada and M. Matsuda, 3–19. Tokyo: Yuhikaku.

———. 2005. "Youth Culture and the Shaping of Japanese Mobile Phones." In *Personal, Portable, Pedestrian: Mobile Phones in Japanese Life*, ed. M. Ito, D. Okabe, and M. Matsuda, 41–60. Cambridge, MA: MIT Press.

Okin, J. R. 2005. *The Information Revolution: The Not-For-Dummies Guide to the History, Technology, and Use of the World Wide Web*. Winter Harbor, ME: Ironbound Press.

Owyang, Jeremiah. 2006. "Early Bloggers (Before the Internet) at Speaker's [*sic*] Corner, London." Jeremiah the Web Prophet, posting for February 27. Available at http://jeremiahthewebprophet.blogspot.com/2006/02/early-bloggers-before-internet-at.html.

Page, Benjamin, and Jason Tannenbaum. 1996. "Populistic Deliberation and Talk Radio." *Journal of Communication* 46 (2): 33–54.

Palander-Collin, Minna. 1999. "Male and Female Styles in Seventeenth Century Correspondence." *Language Variation and Change* 11:123–141.

Paquet, Sébastien. 2002. "Personal Knowledge Publishing and Its Uses in Research." Seb's Open Research, October 1. Available at http://radio.weblogs.com/0110772/stories/2002/10/03/personalKnowledgePublishingAndItsUsesInResearch.html.

Parker, Ian. 2001. "Absolute PowerPoint." *New Yorker*, May 28, 76.

Parkes, M. B. 1991. *Scribes, Scripts, and Readers: Studies in the Communication, Presentation, and Dissemination of Medieval Texts*. London: Hambledon Press.

Perry, Charles R. 1977. "The British Experience 1876–1912: The Impact of the Telephone During the Years of Delay." In *The Social Impact of the Telephone*, ed. I. Pool, 69–96. Cambridge, MA: MIT Press.

Peters, Andrew. 2006. "Facebook Filled with Interest and Ignorance." *Tartan: Carnegie Mellon's Student Newspaper*, April 3.

Pew Research Center for the People and the Press. 2005. "Public More Critical of Press, But Goodwill Persists." Survey conducted in association with the Project for Excellence in Journalism, June 26. Available at http://people-press.org/reports/display.php3?ReportID=248.

Plotz, David. 2002. "The Plagiarist: Why Stephen Ambrose is a Vampire." *Slate*. Available at http://www.slate.com/?id=2060618.

Pool, Marina M., Cees M. Koolstra, and Tom H.A. Van der Voort. 2003. "Distraction Effects of Background Soap Operas on Homework Performance." *Educational Psychology* 23:361–380.

Pressler, Margaret Webb. 2007. "For Texting Teens, an OMG Moment When the Phone Bill Arrives." *Washington Post*, May 20, A1.

Putnam, Robert. 2000. *Bowling Alone: The Collapse and Revival of American Community*. New York: Simon & Schuster.

Quan-Haase, Anabel, and Barry Wellman. 2002. "Capitalizing on the Net: Social Contact, Civic Engagement, and Sense of Community." In *The Internet in Everyday Life*, ed. B. Wellman and C. Haythornthwaite, 291–324. Malden, MA: Blackwell.

Ralli, Tania. 2005. "Brand Blogs Capture the Attention of Some Companies." *New York Times*, October 24, C6.

Rheingold, Howard. 1993. *The Virtual Community: Homesteading on the Electronic Frontier*. Reading, MA: Addison Wesley.

———. 1999. "Look Who's Talking." *Wired* 7.01, January. Available at http://www.wired.com/wired/archive/7.01/amish_pr.html.

Richtel, Matt. 2007a. "It Don't Mean a Thing if You Ain't Got That Ping." *New York Times*, April 22, Week in Review, 5.

———. 2007b. "Hands on the Wheel, Not on the BlackBerry Keys." *New York Times*, May 12.

Rideout, Victoria, Donald F. Roberts, and Ulla G. Foehr. 2005. "Generation M: Media in the Lives of 8-18 Year-Olds." Kaiser Family Foundation, March. Available at http://www.kff.org/entmedia/entmedia030905pkg.cfm.

Riesman, David. 1950. *The Lonely Crowd: A Study of the Changing American Character*. New Haven, CT: Yale University Press.

Rimer, Sara. 2007. "For Girls, It's Be Yourself, and Be Perfect, Too." *New York Times*, April 1, A1.

Rivière, Carole, and Christian Licoppe. 2005. "From Voice to Text: Continuity and Change in the Use of Mobile Phones in France and Japan." In *The Inside Text*, ed. R. Harper, L. Palen, and A. Taylor, 103–126. Dordrecht: Springer.

Robertson, Jean. 1942. *The Art of Letter Writing: An Essay on the Handbooks Published in England during the Sixteenth and Seventeenth Centuries*. London: Hodder & Stoughton Ltd.

Rodriguez, Felix, and Garland Cannon. 1994. "Remarks on the Origin and Evolution of Abbreviations and Acronyms." In *English Historical Linguistics, Papers from the 7th International Conference on English Historical Linguistics*, ed. F. Fernandez, M. Fuster, and J. Calvo, 261–272. Amsterdam: John Benjamins.

Rogers, Robert, and Stephen Monsell. 1995. "Costs of a Predictable Switch between Simple Cognitive Tasks." *Journal of Experimental Psychology—General* 124:207–231.

Romaine, Suzanne. 2003. "Variation in Language and Gender." In *The Handbook of Language and Gender*, ed. J. Holmes and M. Meyerhoff, 98–118. Malden, MA: Blackwell.

Rose, Mark. 1993. *Authors and Owners: The Invention of Copyright*. Cambridge, MA: Harvard University Press.

Rosenberg, Scott. 1995. "Lost Highway: Tripping Down Bill Gates' Road to Nowhere." *Salon*, December 2. Available at http://www.salon.com/02dec1995/departments/gates.html.

Rosenzweig, Roy. 2006. "Can History Be Open Source? Wikipedia and the Future of the Past." *Journal of American History* 93 (1):117–146.

Rubenstein, Joshua S., David E. Meyer, and Jeffrey E. Evans. 2001. "Executive Control of Cognitive Processes in Task Switching." *Journal of Experimental Psychology—Human Perception and Performance* 27:763–797.

Ruuskanen, Olli-Pekka. 2004. "Essay 4. More than Two Hands: Is Multitasking an Answer to Stress?" *An Economic Analysis of Time Use in Finnish Households*. PhD dissertation, Helsinki School of Economics, Helsinki, Finland, 188–229.

Sacks, Harvey, Emanuel Schegloff, and Gail Jefferson. 1974. "A Simplest Systematics for the Organization of Turn-Taking in Conversation." *Language* 50:696–735.

Samuelson, Robert. 2006. "A Web of Exhibitionists." Op-Ed, *Washington Post*, September 20.

Sanger, Larry. 2004. "Why Wikipedia Must Jettison Its Anti-Elitism." Kuro5hin, December 31. Available at http://www.kuro5hin.org/story/2004/12/30/142458/25.

———. 2006. "The Early History of Nupedia and Wikipedia: A Memoir." In *Open Sources 2.0.*, ed. C. DiBona, D. Cooper, and M. Stone, 307–338. Sebastapol, CA: O'Reilly. Available at http://features.slashdot.org/article.pl?sid=05/04/18/164213&tid=95.

Sapir, Edward. 1921. *Language: An Introduction to the Study of Speech*. New York: Harcourt, Brace.

Saulny, Susan. 2002. "Convictions and Charges Voided in '89 Central Park Jogger Attack." *New York Times*, December 20, A1

Saunders, J. W. 1951. "The Stigma of Print: A Note on the Social Basis of Tudor Poetry." *Essays in Criticism* 1:139–164.

Saunders, W. M. 1921. *The Motor Trade Telegram Code*. London: The Society of Motor Manufacturers and Traders, Limited.

Schegloff, Emanuel, and Harvey Sacks. 1973. "Opening up Closings." *Semiotica* 8:289–327.

Schiano, Diane, Coreena Chen, Jeremy Ginsberg, Unnur Gretarsdottir, Megan Huddleston, Ellen Isaacs. 2002. "Teen Use of Messaging Media." *Proceedings of the ACM Conference on Human Factors in Computing Systems* (CHI '02). Minneapolis, MN, April 20–25. New York: ACM Press, 594–595.

Schiano, Diane, Bonnie Nardi, Michelle Gumbrecht, and Luke Swartz. 2004. "Blogging by the Rest of Us." *Proceedings of the ACM Conference on Human Factors in Computing Systems* (CHI '04). Vienna, Austria, April 24–29. New York: ACM Press, 1143–1146.

"Schools Say iPods Become Tool for Cheaters." 2007. Associated Press, April 27. Available at *CNN News* at http://www.cnn.com/2007/EDUCATION/04/27/ipod.cheating.ap/index.html?eref=rss_topstories.

sciam.com. 2004. "*Scientific American Mind* Poll: 90% of American Adults are Multitaskers." Press release summarizing results of *Scientific American Mind* / Harris Interactive study, December 20. Available at http://pr.sciam.com/release.cfm?site=sciammind&date=2004-12-20.

Scott, Gini Graham. 1996. *Can We Talk? The Power and Influence of Talk Shows*. New York: Plenum Press.

Scott, William, and Jarvogen, Milton. 1868. *A Treatise upon the Law of Telegraphs*. Boston: Little, Brown and Company.

Selgin, George. 2003. "Gresham's Law." *EH.Net Encyclopedia*, ed. R. Whaples, June 10. Available at http://eh.net/encyclopedia/article/selgin.gresham.law.

Sellen, Abigail J., and Richard Harper. 2002. *The Myth of the Paperless Office*. Cambridge, MA: MIT Press.

Serfaty, Viviane. 2004. *The Mirror and the Veil: An Overview of American Online Diaries and Blogs*. Amsterdam: Rodopi.

Sharpe, J. A. 1985. "Last Dying Speeches: Religion, Ideology, and Public Execution in Seventeenth-Century England." *Past and Present* 107:144–167.

Shaw, George Bernard. 1985. *Agitations: Letters to the Press 1875–1950*. Edited by Dan H. Laurence and James Rambeau. New York: Frederick Ungar.

Shea, Virginia. 1994. *Netiquette*. San Francisco: Albion Books.

Sheidlower, Jesse. 1996. "Elegant Variation and All That." Review of *The New Fowler's Modern English Usage* by H. W. Fowler, ed. R. W. Burchfield. New York: Oxford University Press. *Atlantic Monthly* 278 (December): 112–117.

Sherbo, Arthur. 1997. *Letters to Mr. Urban of the Gentleman's Magazine, 1751–1811*. Lewiston, NY: Edwin Mellen Press.

Shipley, David, and Will Schwalbe. 2007. *Send: The Essential Guide to Email for Home and Office*. New York: Alfred Knopf.

Shiu, Eulynn, and Amanda Lenhart. 2004. "How Americans Use Instant Messaging." Pew Internet & American Life Project, September 1. Available at http://www.pewinternet.org/PPF/r/133/report_display.asp.

Silverstone, Roger, and Leslie Haddon. 1996. "Design and Domestication of Information and Communication Technologies: Technical Change and Everyday Life." In *Communication by Design: The Politics of Information and Communication Technologies*, ed. R. Silverstone and R. Mansell, 44–74. Oxford: Oxford University Press.

Sipress, Alan. 2006. "Where Real Money Meets Virtual Reality, the Jury is Still Out." *Washington Post*, December 26, A1.

Smith, Anthony. 1979. *The Newspaper: An International History*. London: Thames and Hudson.

"Some Tech-Gen Youth Go Offline." 2006. Associated Press, October 6. Available at *Wired News* at http://wired.com/news/wireservice/1,71918-0.html.

Speakers' Corner. Available at http://www.speakerscorner.net.

Sprint. 2004. "Sprint Survey Finds Nearly Two-Thirds of Americans are Uncomfortable Overhearing Wireless Conversations in Public." News release, July 7. Available at http://www2.sprint.com/mr/news_dtl.do?id=2073.

Sproull, Lee, and Sara Kiesler. 1986. "Reducing Social Context Cues: Electronic Mail in Organizational Communication." *Management Science* 32:1492–1512.

———, eds. 1987. *Computing and Change on Campus*. New York: Cambridge University Press.

————. 1991. *Connections: New Ways of Working in the Networked Organization.* Cambridge, MA: MIT Press.

Squires, Lauren. 2007. "Whats the Use of Apostrophes? Gender Difference and Linguistic Variation in Instant Messaging." *American University TESOL Working Papers*, no. 4. Available at http://www.american.edu/tesol/WorkingPapers04.html.

Stone, Brad. 2006. "Web of Risks." *Newsweek*, August 21–28, 76–77.

————. 2007a. "Bereft of Blackberries, the Untethered Make Do." *New York Times*, April 19, C1.

————. 2007b. "Facebook Expands into MySpace's Territory." *New York Times*, May 25, C1.

Stroop, John R. 1935. "Studies of Interference in Serial Verbal Reactions." *Journal of Experimental Psychology* 18:643–662.

Stutzman, Frederic. 2006. "An Evaluation of Identity-Sharing Behavior in Social Network Communities." *International Digital and Media Arts Journal* 3 (1). Available at http://www.idmaa.org/journal/archive.htm.

Tagliamonte, Sali, and Derek Denis. 2006. "LOL for Real! Instant Messaging in Toronto Teens." Paper presented at the Linguistic Association of Canada and the United States (LACUS). Toronto, Canada, July 31–August 4.

Tannen, Deborah. 1982a. "Oral and Written Strategies in Spoken and Written Narratives." *Language* 58:1–21.

————. 1982b. "The Oral/Literate Continuum in Discourse." In *Spoken and Written Language: Exploring Orality and Literacy*, ed. D. Tannen, 1–16. Norwood, NJ: Ablex.

————. 1994. *Gender and Discourse*. New York: Oxford University Press.

————. 2003. "Did You Catch That? Why They're Talking as Fast as They Can." *Washington Post*, January 5, B1, B4.

"Taxi Drivers' Brains 'Grow' on the Job." 2000. *BBC News*, March 14. Available at http://news.bbc.co.uk/1/hi/sci/tech/677202.stm.

Taylor, Alex S., and Richard Harper. 2002. "Age-Old Practices in the 'New' World: A Study of Gift-Giving between Teenage Mobile Phone Users." *Proceedings of the ACM Conference on Human Factors in Computing Systems* (CHI '02). Minneapolis, MN, April 20–25. New York: ACM Press, 439–446.

"Telecommunications and Information Highways: Global—Internet—Statistics Overview." 2006. Report from Paul Budde Communication Pty Ltd.

"The Telegraph and the Mails." 1874. *Telegrapher* 7 (February 16): 57.

Thomson, Rob, Tamar Murachver, and James Green. 2001. "Where is the Gender in Gendered Language?" *Psychological Science* 12:171–175.

Thurlow, Crispin. 2006. "From Statistical Panic to Moral Panic: The Metadiscursive Construction and Popular Exaggeration of New Media Language in the Print Media." *Journal of Computer-Mediated Communication* 11(3). Available at http://jcmc.indiana.edu/vol11/issue3/thurlow.html.

Thurlow, Crispin, with Alex Brown. 2003. "Generation Txt? The Sociolinguistics of Young People's Text-Messaging." *Discourse Analysis Online*. Available at http://extra.shu.ac.uk/daol/articles/v1/n1/a3/thurlow2002003-paper.html.

Transactions of the Book. 2001. A Conference at the Folger Shakespeare Library, Washington, DC, November 1–3.

Traugott, Michael, Sung-Hee Joos, Rich Ling, and Ying Qian. 2006. *On the Move: The Role of Cellular Communication in American Life*. Pohs Report on Mobile Communication. Ann Arbor: University of Michigan.

Tufte, Edward. 2003. *The Cognitive Style of PowerPoint*. Cheshire, CT: Graphics Press.

Turkle, Sherry. In press. "Always-On/Always-On-You: The Tethered Self." In *Handbook of Mobile Communication Studies*, ed. J. Katz. Cambridge, MA: MIT Press.

Turner, Fred. 2006. *From Counterculture to Cyberculture: Stewart Brand, the Whole Earth Network, and the Rise of Digital Utopianism*. Chicago: University of Chicago Press.

Turow, Joseph. 1974. "Talk Show Radio as Interpersonal Communication." *Journal of Broadcasting* 18 (2): 171–179.

Ullman, B. L. 1960. *The Origin and Development of Humanistic Script*. Rome: Edizioni di Storia e Letteratura.

Umble, Diane Zimmerman. 1996. *Holding the Line: The Telephone in Old Order Mennonite and Amish life*. Baltimore: Johns Hopkins University Press.

"US Highway Deaths on the Rise." 2006. Associated Press, August 22. Available at *MSNBC* at http://www.msnbc.msn.com/it/14470457/print/1/displaymode/1098/.

"USA SMS Traffic Almost Doubled in 2006." 2007. *Cellular-News*, March 29. Available at http://www.cellular-news.com/story/22869.php.

Vanden Boogart, Matthew Robert. 2006. *Discovering the Social Impacts of Facebook on a College Campus*. Master's thesis, Department of Counseling and Educational Psychology, College of Education, Kansas State University, Manhattan, Kansas.

Van Doren, Charles. 1962. "The Idea of an Encyclopedia." *American Behavioral Scientist* 6(1):23–26.

Veenhof, Ben. 2006. "The Internet: Is It Changing the Way Canadians Spend Their Time?" Science, Innovation and Electronic Information Division, Statistics Canada, August. Available at http://www.statcan.ca.

Vershbow, Ben. 2006. "The Networked Book." Forbes.com, December 1. Available at http://www.forbes.com/2006/11/30/future-books-publishing-tech-media_cz_bv_books06_1201network_print.html.

Watkins, Erin. 2007. "Instant Fame: Message Boards, Mobile Phones, and Clay Aiken." *American University TESOL Working Papers*, no. 4. Available at http://www.american.edu/tesol/WorkingPapers04.html.

Weedon, Alexis. 2003. *Victorian Publishing: The Economics of Book Production for a Mass Market, 1836–1916*. Aldershot, England: Ashgate.

Weeks, Linton. 2001. "The No-Book Report: Skim It and Weep." *Washington Post*, May 14, C1.

Wellman, Barry, Anabel Quan, James Witte, and Keith Hampton. 2001. "Does the Internet Increase, Decrease, or Supplement Social Capital? Social Networks, Participation, and Community Commitment." *American Behavioral Scientist* 45: 436–455.

Westin, Alan F. 1967. *Privacy and Freedom.* New York: Atheneum.

Woodmansee, Martha. 1984. "The Genius and the Copyright: Economic and Legal Conditions of the Emergence of the 'Author.'" *Eighteenth-Centuries Studies* 17: 425–448.

Woodmansee, Martha, and Peter Jaszi, eds. 1994. *The Construction of Authorship: Textual Appropriation in Law and Literature.* Durham, NC: Duke University Press.

Wright, Robert. 2007. "E-Mail and Prozac." Op-Ed, *New York Times,* April 17.

Wurtzel, Alan H., and Colin Turner. 1977. "What Missing the Telephone Means." *Journal of Communication* 27 (2): 48–57.

Yates, Simeon. 1996. "Oral and Written Aspects of Computer Conferencing." In *Computer-Mediated Communication: Linguistic, Social, and Cross-Cultural Perspectives,* ed. S. Herring, 22–46. Amsterdam: John Benjamins.

Zernike, Kate. 2002. "Book-Club Smarts in a Nutshell: Get Notes." *New York Times,* July 19, A1.

Index